Increase your efficiency. Increase your score.

Advanced Math Workbook for the SAT

3rd Edition

Roland Kim, Ph.D.

Living Free Publishing Co.

About the author:

Roland Kim holds a M.A. and Ph.D. (1999) in clinical psychology from Rosemead School of Psychology as well as a M.A. in economics from University of California in Los Angeles and University of Hawaii. He is a four-time perfect scorer (800) in the math section of the Graduate Record Exam and has spent 15 years teaching and advising hundreds of college-bound students, helping them to gain admission into highly competitive universities, including IVY League colleges.

ADVANCED MATH WORKBOOK for the SAT
Living Free Publishing Co.
www.advancedmathworkbook.com

Printed in the United States of America

Third Edition; 2021; Second edition: 2011; First edition: 1998

Please refer to the back of the book for ordering information.

TO THE READER

After advising hundreds of college-bound students for the past 15 years having taken the GRE several times myself, I have gained important insights into preparing for the mathematics sections of the SAT, GRE and GMAT. I have especially found that these tests are designed to not only assess test-takers' abilities to solve certain problems, but also their capacities to approach them efficiently enough to finish the entire problem set within the time limit. Most test prep workbooks available, however, predominantly focus on simply finding the answer rather than illustrating how to efficiently approach the problems. As a four-time perfect scorer in the math section of the GRE, I have concluded that the reason I was able to finish the test so quickly (within 10 to 15 minutes) was that I was able to efficiently and accurately approach each problem.

This book is unique in several respects. First, as opposed to serving as an introductory exam prep book or a math refresher course, it is intended for those students who are motivated to raise their scores from a 500 to an 800. Therefore, as it is assumed that these students are already familiar with exam format, basic algebra and arithmetic, I have skipped these explanations in some sections of the book. Second, in contrast to the hit-and-miss strategies presented in traditional exam prep courses or workbooks, my perspective-oriented approaches can help students improve their test performance by augmenting their problem-solving speed and accuracy. You will understand how to solve more advanced algebra, word and geometry problems through my efficient "Bird's-Eye View" approach as well as many other of my innovative methods. Finally, these methods are designed help test-takers maintain confidence in their abilities to solve these problems. From my own experience, I have found that strategies such as backward-solving, plugging in numbers or guessing—though they yield (limited) success—actually slow down the problem-solving process and at their worst, diminish student motivation to achieve a feeling of mastery and aim for a perfect score.

A comprehensive diagnostic test and the Math Analysis sheet will help you to locate your weaker areas in math preparation for the SAT, GRE or GMAT. Once you identify your problem area(s), you are given a chance to try some sample questions to identify your specific weakness or inefficiency in that particular type of problem. After this initial attempt, you will be given a concise illustration on how to approach these problems more efficiently. The practice questions and chapter tests that follow are challenging enough for you to learn these much more efficient approaches by trying them out only once.

In preparation for the first edition of this book, published in 1998, I extend my thanks to those students who had spent their time with me, as well as their parents who supported them spending time with me. I am also grateful to my colleagues, Dr. Mark Bernstein, English teacher at Flintridge Sacred Heart Academy in La Cañada (Flintridge, CA) and Lydia Cho, my former student and a graduate of Pomona College, for their invaluable suggestions and laborious proofreading of the manuscript.

In preparing for the revised second editions of this book for the SAT, GRE, and GMAT, I greatly owe thanks to my three children. I especially my family for editing the content and redesigning the layout of the book, creating a new cover and spending many hours on the answer key, and helping me prepare the website, www.advancedmathworkbook.com.

I hope this book will enable many motivated students to pursue their academic and professional careers successfully.

Roland Kim, June 2021

HOW TO USE THIS BOOK

There are two different types of problem-solving questions presented in this book: multiple-choice and numeric entry without choices.

Multiple-choice questions require you to select the one best answer among five choices, (A) to (E).

Numeric entry questions require you to produce a numeric answer directly as an integer, decimal or fraction in grid answer boxes.

If you find difficulty in solving a SAT mathematics problem,

Step 1: Identify what **type of problem** it is according to the method used in this book. In the diagnostic test, each question represents the type of problems on which you will want to train yourself. These numbers are cross-referenced to the corresponding section in the book in the Math Analysis Sheet.

Step 2: Without looking at the instructions, complete the **sample questions** within the alloted time. Through this, you will be able to identify your weakness in handling this type of problem.

Step 3: Read and familiarize yourself with the **"general rules" and instructional content** following the sample problem(s).

Step 4: Study the **answers and approaches** to these sample questions, which are available after the general rules, and compare them with your own.

Step 5: Try the **practice questions** following the sample approaches and answers.

Step 6: Compare your approaches with what is demonstrated in the **answer key** located in the back of the book. You may find yourself performing many unnecessary calculations, but you can train yourself to find the ways, as demonstrated, to answer these questions more efficiently.

REMEMBER: You won't be satisfied simply by solving the probem. Rather, it is how efficiently you deal with the problem that makes the dramatic difference in your score!

TABLE OF CONTENTS

To the Reader / How to Use This Book .... 3 / 5
Diagnostic Test / Answers .... 8 / 20
Math Analysis Sheet 21

Chapter 1 :: Bird's-Eye View Approach to Algebra
1.1 Factor Out! 24
1.2 Add/Subtract the Whole 27
1.3 Multiply/Divide the Whole 29
1.4 Break up or Reunite 32
1.5 Compare & Transform 34
1.6 Don't Calculate but Write Out 37
Chapter 1 Test 39

Chapter 2 :: Numbers
2.1 Integers - Positive/0/Negative Integers 44
2.2 Integers - Even & Odd Integers / Multiples & Divisibility 46
2.3 Prime Factorization 49
2.4 Fractions & Decimals - L.C.D. 51
2.5 Min/Max Integer Search 53
2.6 Digit Search 55
Chapter 2 Test 58

Chapter 3 :: Fundamental Rules of Algebraic Operation
3.1 Fundamental Rules of Arithmetic & Basic Algebra: Overview 62
3.2 Polynomial Product & Factoring 66
3.3 Finding a Variable in Terms of Others 69
3.4 Operations in Inequality 72
3.5 Exponential Operations 76
3.6 Roots & Radical Operations 79
3.7 Linear Functions 82
3.8 General Functions & Graphs 86
3.9 Quadratic / Higher Order Functions, Graphs, & Equations 89
3.10 Absolute Value Functions, Equations, & Inequalities 93
Chapter 3 Test 97

Chapter 4 :: Special Types of Algebraic Problems
4.1 Extra-Terrestrial 102
4.2 Division & Remainder 105
4.3 Sequence & Series / Pattern Search 108

4.4 Logic/Set/Counting Problem 111
4.5 Number of Cases / Probability 114
4.6 Statistics - Mean, Median, & Mode 119
4.7 Statistics - Data Interpretation 122
Chapter 4 Test 129

Chapter 5 :: Word Problems

5.1 Conversion Formulas 134
5.2 One-Variable Approach 136
5.3 Consecutive Integers 139
5.4 Digit 142
5.5 Age 145
5.6 Proportion 148
5.7 Ratio 151
5.8 Average 154
5.9 Percent 157
5.10 Motion 161
5.11 Mixture 164
5.12 Work 168
Chapter 5 Test 171

Chapter 6 :: Special Types of Word Problems

6.1 Set (Venn Diagram) 178
6.2 Fractions in Word Problems 181
6.3 Utility Bill / Taxi Fare 184
6.4 Cost/Production Comparison 187
Chapter 6 Test 189

Chapter 7 :: Geometry

7.1 Angle Relations 192
7.2 Parallel Lines 195
7.3 Isosceles, Equilateral, & Right Triangles 198
7.4 Special Right Triangles / Pythagorean Theorem 201
7.5 Inequality in Triangles 204
7.6 Perimeter & Circumference 207
7.7 Polygons - Features/Area 210
7.8 Bird's-Eye View in Area/Perimeter 214
7.9 Circles & Related Topics 217
7.10 Surface Area / Volume / Similar Figures 221
7.11 Coordinate Geometry 225
7.12 Counting & Min/Max Search 228
7.13 Algebra in Geometry 231
7.14 Test of Imagination / Drawing 234
Chapter 7 Test 237

Answer Key 241

DIAGNOSTIC TEST

Please test yourself against time and compute your **Problem Solving Efficiency**.

Number of correct answers (maximum 60): Number wrong or not attempted: ( )	________	problems
	÷	
Total time taken finish the test (maximum 60 minutes):	________	minute
Your **Problem Solving Efficiency** (/1; minimum 0): (Total number of correct answers ÷ number of minutes taken)	________	problems / minute

Each question represents a section from each chapter of this book. If you spend more than 30 seconds on each question (Problem Solving Efficiency < 2), you have some room to improve. You have much more room to improve if you spend more than one minute oneach question (Problem Solving Efficiency < 1).

1 $\frac{11^5 - 11^4}{10}$

(A) 1/10 (B) 11/10 (C) $11^3/10$ (D) $11^4/10$ (E) 11^4

2 a + b = 5, b + c = 12, and c + a = 13. What is a + b + c?

(A) 13
(B) 15
(C) 17
(D) 19
(E) 20

3 ab = 3, bc = 6, ca = 2, and a > 0. What is abc?

(A) 3
(B) 4
(C) 6
(D) 8
(E) 12

If x and y are integers and $\frac{(x^2 - y^2)}{xy} = 2$, what is $\frac{x}{y} - \frac{y}{x}$?

(A) 1/2
(B) 2
(C) 6
(D) 8
(E) 12

What is k if $\frac{3.5 + 0.125}{2.5} = \frac{70 + 2.5}{k}$?

(A) 37.5
(B) 45
(C) 50
(D) 60
(E) 75

y denotes the sum of the odd integers from 1 to 49 inclusive, and x denotes the sum of the odd integers from 51 to 99 inclusive. What is the value of x - y?

(A) 500
(B) 600
(C) 750
(D) 1,000
(E) 1,250

If x is not an integer, which of the following can be an integer?

(A) x/2 (B) x^2 (C) $\sqrt{x-1}$ (D) x + 1 (E) 1/x

T is a 5-digit number with 5, 0, and 0 as its last three digits, in that order. T can therefore be divisible by each of the following EXCEPT

(A) 8 (B) 25 (C) 100 (D) 125 (E) 250

9 How many prime numbers are there between 15 and 30?

(A) 3 (B) 4 (C) 5 (D) 6 (E) 7

10 Ms. Green owns 2/3 of the shares in a company. If she sells 3/4 of her shares, what fraction of the shares does she own now?

(A) 1/12
(B) 5/12
(C) 1/6
(D) 1/5
(E) 1/4

11 The average of three different positive integers is 5. What is the greatest possible value of the product of the three integers?

(A) 56
(B) 84
(C) 100
(D) 120
(E) 150

12 In the correctly-worked addition problem below, which of the following could be the digit A?

$$\begin{array}{r} AB \\ +\ \ 6D \\ \hline 152 \end{array}$$

I. 7
II. 8
III. 9

(A) II only
(B) III only
(C) I and II only
(D) II and III only
(E) I, II, and III

13 $\frac{x}{y}+\frac{y}{x}=8$. What is the value of $\dfrac{x+y}{\frac{1}{x}+\frac{1}{y}}$?

(A) 5
(B) 8
(C) 10
(D) 12
(E) 16

14 $x^2 + y^2 = 44$, and $xy = 4$. What is $(x - y)^2$?

(A) 15
(B) 24
(C) 25
(D) 30
(E) 36

15 If $\frac{3}{2x - 1} = y$, what is x in terms of y?

(A) $\frac{y - 3}{2y}$ (B) $\frac{2y}{y - 3}$ (C) $\frac{y}{2y - 3}$ (D) $\frac{y + 3}{2y - 3}$ (E) $\frac{y}{y - 3}$

16 If $5 < x < 8$ and $11 < y < 15$, then what is $1/(y - x)$ between?

(A) 1/10 and 1
(B) 1/3 and 1
(C) 1/5 and 1/3
(D) 1/10 and 1/5
(E) 1/10 and 1/3

17 $5^{20} \bullet 6 - 5^{21} =$

(A) 5^{20}
(B) 5^{27}
(C) 5^{140}
(D) 35^{20}
(E) 35^{140}

18 x and y are both negative. What can $\sqrt{x^2 + y^2}$ NOT be?

(A) 0
(B) Negative
(C) Positive
(D) Both A and B
(E) Both A and C

19 If line T is the graph of the equation $3x + 4y = 5$ and the point at which T crosses the x-axis has coordinates (h, 0), what is the value of h?

(A) 0
(B) 5/3
(C) 7/3
(D) 8/3
(E) 11/2

20 If $f(x) = x - 1$ and $g(x) = x^2 - 1$, which of the following pairs of x-values satisfy the equation $f[g(x)] = 0$?

(A) -2, 1
(B) -1, 0
(C) 0, 1
(D) 0, 2
(E) 1, 2

21 If $y = -x^2 + 1$ intersects line k at (t, 3) and (p, 0), what is the maximum possible slope for line k?

(A) -3
(B) 2
(C) 3
(D) 4
(E) 9

22 Which of the following values of x satisfy the equation $|-x - 2| < |x|$, when $-2 < x < 0$?

(A) -2
(B) -1
(C) 0
(D) 1
(E) 2

23 If P(x) = greatest prime factor less than or equal to x, what is P(49)?

(A) 7
(B) 23
(C) 45
(D) 47
(E) 49

24 S and T are positive integers. If S divided by 5 leaves a remainder of 3, and T divided by 5 leaves a remainder of 4, what is the remainder when S • T is divided by 5?

(A) 0
(B) 1
(C) 2
(D) 5
(E) 6

25 1, 3, 5, 7, 1, 3, 5, 7, ...

The sequence above with the first term, 1, repeats in the pattern 1, 3, 5, 7, indefinitely. What is the sum of the values from the 10th term through the 50th term?

(A) 120
(B) 160
(C) 180
(D) 200
(E) 212

26 How many multiples of 5 are there between 25 and 125, including 25 and 125?

(A) 19
(B) 20
(C) 21
(D) 22
(E) 100

27 What is the probability that the difference between the numbers on the face of two dice thrown is greater than 1?

(A) 5/9
(B) 7/12
(C) 19/36
(D) 13/24
(E) 11/18

28 If x, y and z are the respective mean, median and mode of heights among 10 students and $y < x < z$, which of the following is true about their relationships?

(A) There are more students whose heights are greater than x
(B) The tallest student has a height of z
(C) x is the average of y and z
(D) The shortest student has a height of less than x
(E) The middle student has a height of x

29 According to the data of the pie chart below, 420 students majoring in art are represented by 3x% on the chart. How many students major in liberal arts?

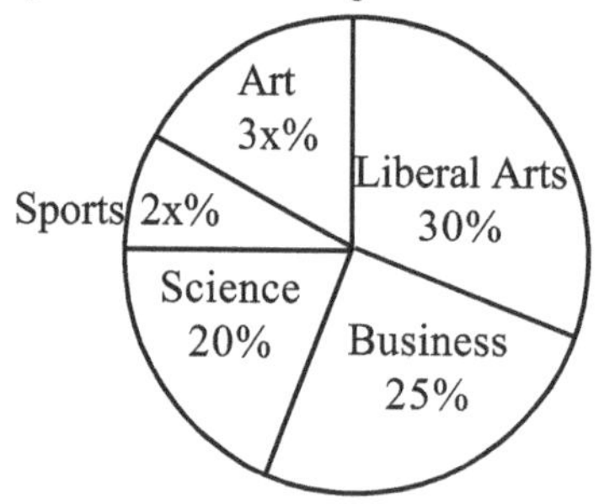

(A) 280
(B) 560
(C) 720
(D) 840
(E) 960

30 If the value of x - 4y is greater than 2x by 300 percent of y, and x and y are not zero, what is x in terms of y?

(A) -7y
(B) -y
(C) y/3
(D) 3y/7
(E) 7y/3

31 In an 25-question test, scores are computed by subtracting 1/4 of the number of incorrect answers from the number of the correct answers. If a student answered every question and scored a 10, how many did he answer incorrectly?

(A) 8
(B) 10
(C) 11
(D) 12
(E) 13

32 If two lists of 4 consecutive even integers are separated by 2 even numbers, the sum of the 4 integers on one list is how much greater than the sum of the other?

(A) 24
(B) 36
(C) 48
(D) 60
(E) Cannot be determined

33 Which of the following cannot be the sum of a two-digit number and the number obtained by reversing the two digits?

(A) 88
(B) 121
(C) 132
(D) 145
(E) 187

34 John's age is twice Nancy's age. 5 years ago, John was x years old. In terms of x, how old will Nancy be in 3 years?

(A) $x + 7$
(B) $(x + 11)/2$
(C) $(x + 2)/3$
(D) $(2x + 13)/4$
(E) $2x + 5$

35 If g tickets cost h dollars, what is the cost of x tickets in cents?

(A) $100h/xg$
(B) $h/100xg$
(C) $100xh/g$
(D) $hg/100x$
(E) xg/h

36 A class is composed of Caucasian, Latino, Asian and African American students in the ratio of 2 : 3 : 1 : 2. If the total number of students in the class is 40, how many students are Latino?

(A) 8
(B) 15
(C) 16
(D) 20
(E) 27

37 If the average (arithmetic mean) of 5 numbers is 4 and the sum of 3 numbers is -16, what is the average of the other two numbers?

(A) -12
(B) -8
(C) 12
(D) 18
(E) 24

38 If y is equal to 5/4 of x, x is what percent of y?

(A) 20%
(B) 33.3%
(C) 80%
(D) 280%
(E) 281%

39 The price of a certain product decreased by 60% every year for the past 4 consecutive years. If the average price of the product was originally \$x, what is the average price of the product now in terms of x?

(A) 16x/625
(B) 16x/125
(C) 16x/81
(D) 36x/81
(E) 18x/9

40 A train traveled 500 miles at x mph and arrived one hour early. The train would have arrived exactly on time if it had traveled at what speed in miles per hour?

(A) x + 1
(B) 500x/(500 + x)
(C) 500/(x + 1)
(D) 500/(500 + x)
(E) x/(x + 500)

41 A bag of candy is made by mixing candy A of \$5 per pound with candy B of \$10 per pound. If the mixture is worth \$7 per pound, how many pounds of candy A are needed to make 300 pounds of the mixture?

(A) 150
(B) 160
(C) 170
(D) 180
(E) 200

42 A 2-pound mixture composed of 1/4 oil and 3/4 water is added to 3 pounds of pure water. What percent of the resulting mixture is water?

(A) 5%
(B) 8%
(C) 10%
(D) 15%
(E) 25%

43 It takes 12 machines 6 days to finish a job. How many days will it take 8 machines, performing at the same rate, to complete 1/3 of the job at the same rate?

(A) 2
(B) 3
(C) 4
(D) 5
(E) 8

44 Ted can finish a job in 3 hours and Ann can finish the same job in 4 hours. If Ted works alone for 1 hour, and then Ann and Ted work together to finish the job, for how many more hours do they have to work together?

(A) 8/7
(B) 6/5
(C) 5/4
(D) 4/3
(E) 3/2

45 Of 70 students taking math, 40 were studying Algebra, 30 were studying Geometry and 10 were studying both. How many students were studying neither?

(A) 0
(B) 4
(C) 10
(D) 23
(E) 30

46 Diane spent 1/4 of her yearly allowance on clothing and spent 2/3 of the remainder on books. If she had $120 left, how much was her allowance?

(A) $240
(B) $360
(C) $400
(D) $480
(E) $600

47 A telephone company charges X dollars for the first 3 minutes of a call and 50 cents for each additional minute, or fraction thereof. If a 20 minute phone call costs $12.00, what is X?

(A) $2.00
(B) $2.50
(C) $3.00
(D) $3.50
(E) $3.75

48 How much more expensive are four pounds of flour selling at $5.00 per 2 pounds than four pounds of flour selling at $7.50 per 2.5 pounds?

(A) $0.50
(B) $0.75
(C) $1.00
(D) $1.50
(E) $2.00

49 In the figure below, what is the value of z in terms of x and y?

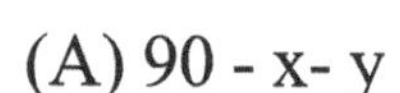
(A) 90 - x- y
(B) 180 - x - 2y
(C) 270 - 2x - 2y
(D) 360 - x- 2y
(E) 540 - 2x - 2y

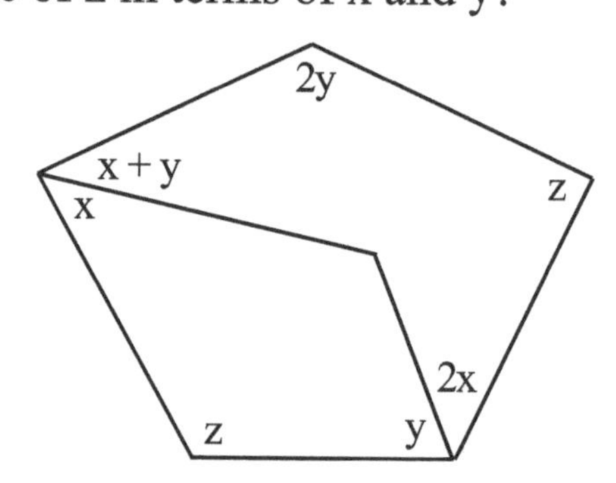

50 In the figure below, if $L_1 // L_2$, what is z in terms of x and y?

(A) 2x - y
(B) 360 - 2x - 2y
(C) 360 - x - y
(D) x - y
(E) y - x

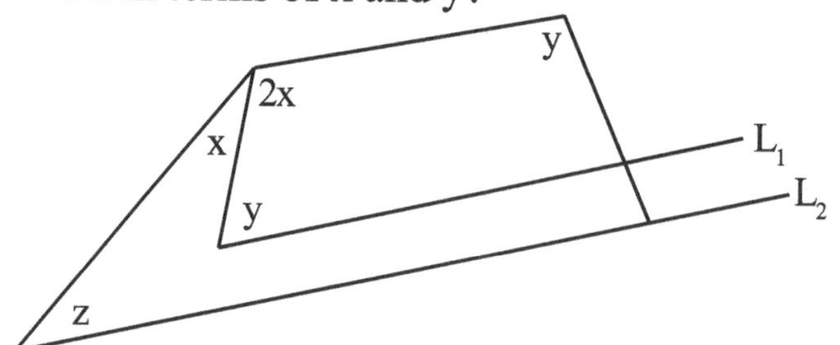

51 If twice the width of a rectangle is equal to 5/6 times the length of the rectangle, what is the ratio of its perimeter to the length of its diagonal?

(A) 17/12
(B) 30/17
(C) 34/15
(D) 30/13
(E) 34/13

52 The area of the right triangle below is $2\sqrt{3}$. What is the length of the altitude h?

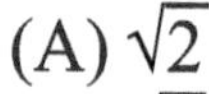
(A) $\sqrt{2}$
(B) $\sqrt{3}$
(C) 2
(D) $2\sqrt{2}$
(E) 3

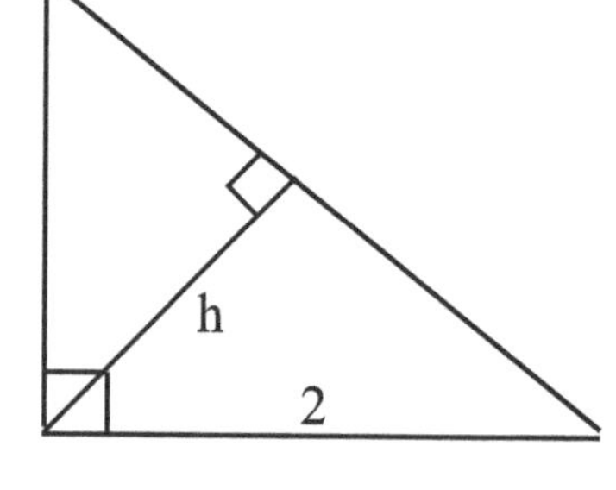

53 In triangle ABC, ∠A is 44° and ∠B is 46°. If D is the point on side AB such that CD⊥AB, which of the following is the shortest?

(A) AC
(B) AD
(C) BC
(D) BD
(E) AD

54 A train is traveling on a circular path with a 1/2 mile radius. If it travels 154 miles, how many laps around this path does it travel? (Use 22/7 for π).

(A) 14
(B) 35
(C) 49
(D) 56
(E) 63

55 If the area of an isosceles right triangle is 9/8 square inches, what is the perimeter of the triangle?

(A) $3 + \sqrt{3}/4$
(B) $3 + \sqrt{3}/2$
(D) $3 + 3\sqrt{2}/2$
(E) $7\sqrt{3}/2$
(C) $4 + 3\sqrt{3}/2$

56 In the figure below, the stair-shaped lines are parallel to either AB or BC. If the length AB is 6 and the length BC is 8, then which of the following represents the sum of line AC and the stair-shaped lines from A to C?

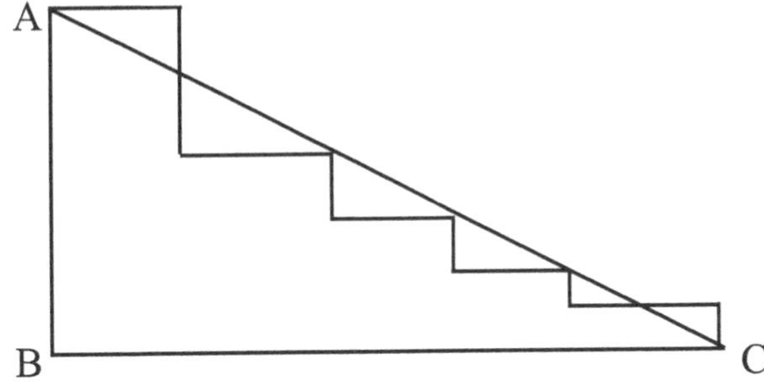

(A) 20
(B) $10 + 10\sqrt{2}$
(C) $10 + 5\sqrt{2}$
(D) 24
(E) $10 + 5\sqrt{3}$

57 Segment AC is the diameter of circle O with the length 10, and AB = 5. Segment BD is the altitude drawn from B to AC. What is the length of BD?

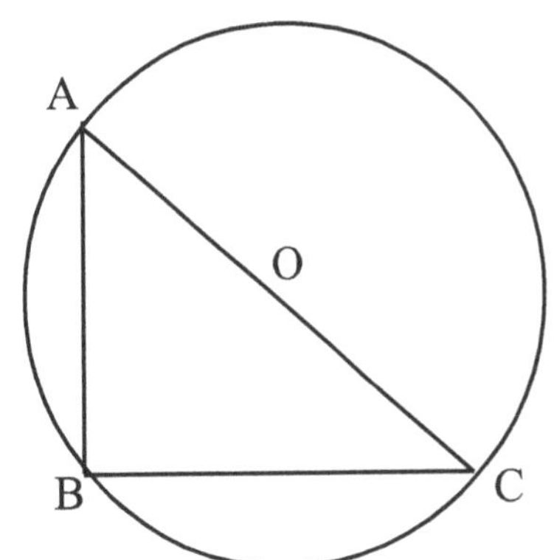

(A) 5
(B) $5\sqrt{3}/2$
(C) $2\sqrt{10}$
(D) $3\sqrt{5}$
(E) $3\sqrt{10}$

58 If the ratio of the surface areas of the two similar cubes in the figure below is 9:25, and if the volume of the smaller cube is 54, what is the volume of the larger cube?

(A) 125
(B) 250
(C) 375
(D) 500
(E) 750

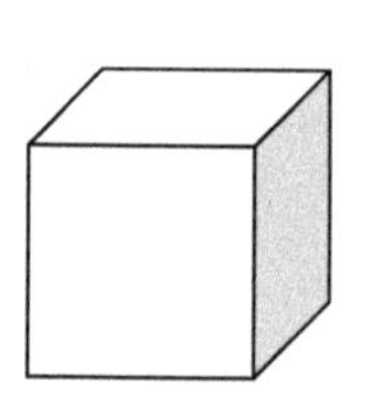
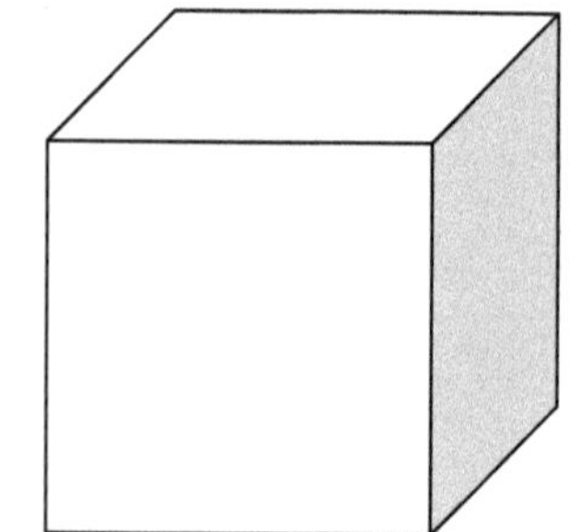

59 If three points A (2, -4), B (0, 4) and C (-3, x) are on the same line, what is the value of x?

(A) -16
(B) -8
(C) 8
(D) 15
(E) 16

60 Four lines can divide a triangular region into a minimum of how many nonoverlapping triangular region(s)?

(A) 0
(B) 1
(C) 2
(D) 3
(E) 4

Answers to Diagnostic Test

1. E	11. D	21. C	31. A	41. D	51. E
2. B	12. D	22. A	32. D	42. C	52. B
3. C	13. C	23. A	33. C	43. B	53. D
4. B	14. E	24. C	34. B	44. A	54. C
5. C	15. C	25. B	35. C	45. C	55. D
6. E	16. E	26. C	36. B	46. D	56. D
7. E	17. A	27. A	37. D	47. D	57. B
8. A	18. D	28. D	38. C	48. E	58. B
9. D	19. B	29. D	39. A	49. C	59. E
10. C	20. D	30. A	40. B	50. E	60. B

MATH ANALYSIS SHEET

Each question in the Diagnostic Test directly corresponds with a section in this book, represented in the table below. Under the column marked "X," mark each test question you incorrectly answered. The table will tell you the problem type, means of identification, and chapter/section through which you can learn to more effectively approach the problem.

For example, if you incorrectly answered question 33 in the Diagnostic Test, you will see that you have room for improvement when it comes to approaching word problems, specifically questions dealing with age. Study Section 5.5, and you will learn to confidently and efficiently solve any problem resembling "5 years ago, John's age was..."

Take note of the key words in the "Identification" column. You already know the basic math; you just have to know when to apply it!

X	#	Problem Type	Identification	Section
Bird's-Eye View				
	1	Factoring	Factorable	1.1
	2	Addition/Subtraction	System of equations	1.2
	3	Multiplication/Division	Can you divide the whole?	1.3
	4	Breaking up/Reuniting	Breakable/combinable	1.4
	5	Compare & Transform	Similar expressions	1.5
Numbers & Arithmetic				
	6	Working without Calculation	Long list of numbers	1.6
	7	Integers/Negative/Positive	Negative/Positive integers	2.1
	8	Even/Odd, Divisibility/Multiples	Even/odd, divisible/multipliable	2.2
	9	Prime Factorization	Prime factors	2.3
	10	Fractions/Decimals	Comparing fractions	2.4
	11	Max/Min Integer Search	Max/min possible integers	2.5
	12	Digit Search	"AB + CD = EFG"	2.6
Algebraic Operations				
	13	Basic Algebra Concept/Equation	Any algebraic operations	3.1
	14	Polynomial Product/Factoring	"$A^2 + b^2 + 2ab = (a + b)^2$"	3.2
	15	Find a Variable in Terms of Others	"x in terms of y and z?"	3.3
	16	Inequality Operation	"$2 < x < 7$"	3.4
	17	Exponents	"$2^n + 2^n =$ or 16^n"	3.5
	18	Roots and Radicals	"Compare 2 and 1.5"	3.6
	19	Linear Functions	"$2x + 3y = 6$" and (x,y)	3.7
	20	General Functions & Graphs	"If the graph f(x)… and g(x)…"	3.8
	21	Quadratic / Higher Order Fn, Eqn.	Quadratic: $ax^2 + bx + c$	3.9
	22	Absolute Value Fn, Eqn, Inequl.	"$y = \|x + 2\|, \|x - 2\| < 0$"	3.10

X	#	Problem Type	Identification	Section
Special Types of Algebraic Problems				
	23	E.T.(Special Symbol)	Strange symbols	4.1
	24	Division/Remainder	"When divided by x, remainder is y"	4.2
	25	Sequence/Pattern Search	"2, 4, 6, ... n"	4.3
	26	Logic/Set/Counting	Logic, "{2, 3, 5, }, how many #'s"	4.4
	27	Number of Cases/Probability	Combinations, probability	4.5
	28	Statistics: Mean, Median, and Mode	Mean, median and mode	4.6
	29	Statistics: Data Interpretation	Charts and graphs	4.7
Word Problems				
	30	Conversion, Simple Word Problems	"2 more than John's...."	5.1
	31	One Variable Approach	"Let x = the unknown, then..."	5.2
	32	Consecutive Integer	Consecutive integers	5.3
	33	Digit word	Tens' and units' digits	5.4
	34	Age	"5 years ago, John's age was ..."	5.5
	35	Proportion	"If a takes b, then c takes...?"	5.6
	36	Ratio	"A : B : C"	5.7
	37	Average	"If the average of x, y and z is..."	5.8
	38	Percent – Direct Conversion	"30 is what % of..."	5.9a
	39	Percent – Increase/Decrease	Increased or decreased by %	5.9b
	40	Motion	Time, rate and distance	5.10
	41	Mixture – Coin/Stamp / Average Price	"How many of each kind…?"	5.11a
	42	Mixture – Solution	% solution	5.11b
	43	Work – Multi-Production	"3 men take 5 days to finish ½ a job"	5.12a
	44	Work – Combined Work	"John and Mary work together"	5.12b
Special Types of Word Problems				
	45	Set Word Problems	"X belongs to A, B, or both groups"	6.1
	46	Fractions in Word Problems	"used 1/3 of the remainder…"	6.2
	47	Utility/Taxi fare	Flat rate of $x and after that	6.3
	48	Cost/Production Comparison	"$4 for 5 products"	6.4
Geometry				
	49	Angle Relations	Angles, triangles, interior angles	7.1
	50	Parallel Lines	Line L_1 // Line L_2	7.2
	51	Types of Triangles	Isosceles and equilateral	7.3
	52	Right Triangle Ratio/Pythagorean	Right triangle and length ratios	7.4
	53	Inequality in Triangles	Side of triangle c > d, angle C > D	7.5
	54	Perimeter/Circumference	Perimeter/ circumference	7.6
	55	Various Polygons/Features & Area	Area of a parallelogram, rectangle	7.7
	56	Bird's-Eye View in Area & Perimeter	"Find the perimeter or area"	7.8
	57	Circle & Related Topics	Anything involving circles	7.9
	58	Surface Area & Volume	Surface area or volume	7.10
	59	Coordinate Geometry	Point A(x, y) and distance AB/slope	7.11
	60	Counting/Min/Max Search: Geometry	"How many regions can be made…"	7.12

Bird's-Eye View Approach to Algebra

1.1 :: Factor Out!

try it yourself Try these four sample questions within 60 seconds.

Q1 If $50^{100} = 2^{100} \bullet 5^{200}x$, what is x?

Q2 If a - b = 5 and $2ab - 2b^2 = 100$, what is b?

Q3 If $x^2 - y^2 = mn$, and x + y = n, what is x - y?

Q4 If $\frac{9^8 - 9^7}{8} = x^7$, then x?

general rules

☑ First, look for **what the question wants.**
☑ Before using the conventional method of variable elimination, try the **"Bird's-Eye View" method of factoring**, utilizing the four essential factoring formulas:

Number factoring: $12 = 2 \bullet 2 \bullet 3 = 2^2 \bullet 3$

Monomial factoring: $2x^2y^2 - 4xy^3 + 2xy = 2xy(xy - 2y^2 + 1)$

Binomial factoring: $4a^2 - 9b^2 = (2a + 3b)(2a - 3b)$

Trinomial factoring: $2x^2 + 3x + 1 = (2x + 1)(x + 1)$

approach to sample questions

A1 Factoring 50^{100} gives us $(2 \bullet 5^2)^{100} = 2^{100} \bullet 5^{200}$. Therefore, $x = 1$.

A2 A simple monomial factoring of $2ab - 2b^2$ yields $2b(a - b) = 100$.
Since $a - b = 5$, dividing both sides by 5 yields $2b = 20$. Therefore, $b = 10$.

A3 Through binomial factoring, $x^2 - y^2 = (x + y)(x - y)$.
Since $x + y = n$, $x - y$ has to be m.

A4 The simple factoring of $9^8 - 9^7$ yields $9^7(9 - 1) = 9^7(8)$.
Therefore, the answer for x is 9.

Practice Questions 1.1

1 If $7(x + y) - 4(x + y) = 15$, then $x + y = ?$

2 If $x + y = 7$ and $x - y = 4$, then $x^2 - y^2 = ?$

3 If $x - y = 6$ and $x^2 - xy = 24$, what is y?

4 Simplify $\sqrt{80(11)^2 + (11)^2}$

5 Simplify $\frac{7^{17} - 7^{16}}{6}$

6 $\frac{10! - 9!}{9!} =$

(A) 10/9
(B) 9
(C) 10
(D) 9!
(E) 10!

1.2 :: Add/Subtract the Whole

try it yourself Try these three sample questions within 30 seconds.

Q1 If $2a - 3b = 5$ and $3a - 2b = 10$, what is $a - b$?

Q2 If

$$x + 2y = 3$$
$$y + 2z = 2$$
$$z + 2x = 7$$

then what is $x + y + z$?

Q3 If

$$4x - 3y = 5$$
$$3x - 4y = 4$$

then what is $x + y$?

general rules

☑ First, look for **what the question wants**. For example, in Q1, what it asks for is not a or b, but (a - b) as a whole.

☑ Try to see the whole picture and **derive the expression** close to what the question asks for by either adding or subtracting.

☑ Before using the conventional method of variable elimination (linear combination or substitution), try the **"Bird's-Eye View" method of addition/subtraction of the whole**.

It is essential that students be familiar with the conventional approaches to solving systems of equations, such as the linear combination or substitution methods, before they learn to apply the Bird's-Eye View method of addition/subtraction of the whole. If the question is asking about an individual variable rather than a whole group of variables, it is better to use the conventional approach.

approach to sample questions

A1 Note that the question asks for $(a - b)$ as a whole rather than a or b separately. Add the two equations to get $5a - 5b = 15$.
By dividing both sides by 5, you get $a - b = 3$.

A2 Note that the question asks for $x + y + z$. Add all the left and right columns respectively to get $3x + 3y + 3z = 12$.
By dividing both sides by 3, you get $x + y + z = 4$.

A3 Note that the question wants $x + y$. By subtracting the bottom equation from the top, you find $x + y = 1$.

Practice Questions 1.2

1 If $3x + 4y = 7$ and $2x + y = 13$, then $x + y = ?$

2 If $4x - 3y = 10$ and $3x - 4y = 5$, then $x + y = ?$

3 If $a + b = 7$
$b + c = 12$
$c + a = 5$ what is the arithmetic mean of a, b, and c?

4 What is the average of x, y, z, and t if
$x + y + z = 18$
$y + z + t = 15$
$x + t = 7$

5 If $4a + b = 22$ and $a + 3b = 23$, then what is $-3a + 2b$?

1.3 :: Multiply/Divide the Whole

try it yourself

Try these three sample questions within 30 seconds.

Q1 If $ab = 3$, $bc = 4$, $ac = 3$ and $a > 0$, what is abc?

Q2 If $x/y = 2/3$ and $y/z = 4/7$, what is x/z?

Q3 If $2(x + y) = 5$ and $5(x + y) = k$, what is k?

general rules

☑ First, look for **what the question wants.**
☑ Before using the conventional method of variable elimination, try the **"Bird's-Eye View" method of multiplying/dividing by the whole.**

approach to sample questions

A1 The question asks for abc rather than a, b, or c separately. If you multiply all three equations together, $ab \bullet bc \bullet ac = 3 \bullet 4 \bullet 3 = 36$.
Therefore, $a^2b^2c^2 = 36$.
Taking the square root, $abc = 6$.

A2 The question asks about x/z. If you multiply x/y by y/z, the y's cancel and the result is x/z. So, $x/z = 2/3 \bullet 4/7 = 8/21$.

A3 Since $x + y = 5/2$, $5(x + y) = 5(5/2) = 25/2$.
Therefore, $k = 25/2$.

Practice Questions 1.3

1 If $mn(x + y) = 10$ and $n(x + y) = 5$, what is m?

2 If $a/b = 4/3$ and $a/c = 3/2$, what is c/b?

3 If $xy^2z^3 = 100$
$yz^2 = 20$ what is xyz?

4 If $3(a - b) = 2k^2$, $5(a - b) = 4k$, and $a \neq b$ and $k \neq 0$, what is k?

5 If $xy = 5$, $yz = 4$, $zx = 5$, and $z > 0$, what is $x^3y^3z^3$?

1.4 :: Break up or Reunite

try it yourself

Try these three sample questions within 30 seconds.

Q1 If x and y are integers and $3 < x < y < 9$, what is the minimum possible value of $\frac{x+y}{xy}$?

Q2 What is $(1 - 1/2) + (1/2 - 1/3) + (1/3 - 1/4) + (1/4 - 1/5)$?

Q3 If $1/a + 1/b = 5$, what is $ab/(a + b)$?

general rules

- ☑ If the expression is in a combined form, **break it up**!
- ☑ If the expression is in a separated form, **combine it**!

approach to sample questions

A1 Breaking $(x + y)/xy$ yields $1/y + 1/x$. Therefore, in order to obtain the smallest possible values of $1/x$ and $1/y$, x and y have to have the minimum values $x = 7$ and $y = 8$. $\frac{7+8}{(7)(8)} = \frac{15}{56}$

A2 When we remove each parenthesis, we will find
$1 - 1/2 + 1/2 - 1/3 + 1/3 - 1/4 + 1/4 - 1/5 = 1 - 1/5 = 4/5$.

A3 Reuniting the two terms, $1/a + 1/b = (a + b)/ab = 5$.
Therefore, $ab/(a + b) = 1/5$.

Practice Questions 1.4

1 If $\frac{x^2 + y^2}{xy} = 10$ and $y/x = 4$, $x/y = ?$

2 What is $(3/4 + 4/5) - (4/5 + 5/6) + (5/6 - 3/4)$?

3 If $\frac{ab + b^2}{b} = 15$ and $b = 7$, what is a?

4 If $a/b + b/a = 15$ and $a^2 + b^2 = 3$, what is ab?

5 If a and b are integers and $4 < a < b \leq 8$, what is the maximum possible value of $\frac{b - a}{ab}$?

1.5 :: Compare & Transform

try it yourself Try these four sample questions within 30 seconds.

Q1 If $2a - 3b = 5$, what is $6b - 4a$?

Q2 What is $\dfrac{1}{y - x} + \dfrac{1}{x - y}$?

Q3 If $\dfrac{(2.5 + 0.175)}{2.2} = \dfrac{(5 + 0.35)}{x}$, then $x = ?$

Q4 If $\dfrac{(2x + y)}{x} = \dfrac{(x - 3y)}{y}$, what is $y/x - x/y$?

general rules

☑ Try to see the **whole picture**.

☑ **Transform** the expression utilizing the rules and properties of inequality:

To change the order, change the sign: $a - b = -(b - a)$

Be fair to both top and bottom of the fraction: $\dfrac{a}{b} = \dfrac{ka}{kb}$

Be fair to both terms in the numerator: $\dfrac{(ka + b)}{a} = k + \dfrac{b}{a}$

approach to sample questions

A1 $6b - 4a = 2(3b - 2a) = -2(2a - 3b) = -2(5) = -10$.

A2 Since $y - x = -(x - y)$, $1/(y - x) = -1/(x - y)$.
$-1(x - y) + 1(x - y) = 0$.

A3 If you multiply the left numerator by 2, the result becomes the right numerator. Therefore, as a rule of proportion, if you multiply the left bottom denominator by 2, it becomes x. So, $x = 4.4$

A4 When you break up the fraction, $2 + y/x = x/y - 3$.
Therefore, $y/x - x/y = -5$.

Practice Questions 1.5

1 Simplify $\dfrac{6a + 2b}{2}$.

2 What is $\dfrac{x - y}{y - x}$?

3 If $5x - 7y = 12$, then what is $14y - 10x$?

4 If $a - 2b = 2t$, what is $\dfrac{6b - 3a}{t}$?

5 If $\dfrac{1}{\frac{1}{x} + 1} = k$, then which of the values below is the same as k?

(A) $\dfrac{1}{\frac{2}{x} + 2}$

(B) $\dfrac{2}{\frac{1}{2x} + 1}$

(C) $\dfrac{2}{\frac{2}{x} + 1}$

(D) $\dfrac{2}{\frac{1}{2x} + \frac{1}{2}}$

(E) $\dfrac{2}{\frac{2}{x} + 2}$

6 Of the following, the closest approximation to $\sqrt{\dfrac{8.98(1589)}{28.99}}$ is

(A) 1
(B) 2
(C) 20
(D) 40
(E) 12

1.6 :: Don't Calculate but Write Out

try it yourself Try these two sample questions within 30 seconds.

Q1 If k is the sum of all the even numbers from 1 to 100 and m is the sum of all the odd numbers from 1 to 100, what is k - m?

Q2 Given that the sum of the odd integers from 1 to 49 is k, what is the sum of the even integers from 2 to 50 in terms of k?

general rules

☑ Try to see the **whole picture**.
☑ Simply **write out** some of the beginning and ending numbers of the sequences or series.
☑ **Add/subtract** the entire list of numbers as a whole.

approach to sample questions

A1 Note that if 2 + 4 + --------------- + 100 = k and
-) 1 + 3 + --------------- + 99 = m
then k - m = 1 + 1 + --------------- + 1 = 50

A2 Note that if 1 + 3 + ------------------ + 49 = k and
+) 1 + 1 + ------------------ + 1 = 25
then 2 + 4 + ------------------ + 50 = k + 25

Practice Questions 1.6

1. Given that the sum of all the multiples of 3 from 3 to 99 is t and the sum of the multiples of 6 from 6 to 198 is s, what is s in terms of t?

2. If the sum of the first 500 positive integers is k, what is the sum of the first 1000 positive integers in terms of k?

3. What is the result when the sum of the odd integers from 1 to 999 inclusive is subtracted from the sum of the even integers from 2 to 1000 inclusive?

4. If the sum of the first n positive integers is $\frac{n(n+1)}{2}$, what is the difference between the sum of the first 100 and the sum of the first 50?

5. If t denotes the sum of the integers from 1 to 250 inclusive, and u denotes the sum of the integers from 251 to 500, what is u - t?

CHAPTER 1 TEST

1 If $x + y = 4$ and $x - y = 2$, then $x^2 - y^2 =$

(A) 4
(B) 8
(C) 12
(D) 16
(E) 64

2 If $2x + 3y = 10$ and $3x + 2y = 10$, then $x + y =$

(A) 2
(B) 3
(C) 4
(D) 5
(E) 6

3 Given that the sum of the odd integers from 1 to 99 is 2,500
(i.e. $1 + 3 + 5 + .. + 99 = 2,500$),
what is the sum of the even integers from 2 to 100?

(A) 2,550
(B) 2,600
(C) 2,750
(D) 3,000
(E) 5,000

4 If $x + y = 12$ and $x^2 + xy = 84$, then $y =$

(A) 5
(B) 6
(C) 7
(D) 8
(E) 10

5 If $3x + y = 15$ and
$x + 2y = 16$, then $2x - y =$

(A) 1
(B) 1/2
(C) 0
(D) -1/2
(E) -1

6 If s denotes the sum of the integers from 1 to 30 inclusive and t represents the sum of the integers from 31 to 60, what is the value of $t - s$?

(A) 30
(B) 300
(C) 450
(D) 900
(E) Cannot be determined

7 If $a - b = 5$, then what is $-(b - a)$?

(A) -5
(B) -3
(C) 5
(D) 8
(E) 10

8 If a and b are positive integers, $a^2 + b^2 = 41$, and $a^2 - b^2 = 9$, then $b =$

(A) 3
(B) 4
(C) 5
(D) 6
(E) 7

9 If $xy = 30$, $yz = 20$, and $zx = 150$, what is xyz?

(A) 20
(B) 45
(C) 120
(D) 240
(E) 300

10 If $3a - 4b = -6$, what is $8b - 6a$?

(A) -12
(B) -8
(C) 10
(D) 12
(E) 24

11 If x denotes the sum of the integers from 1 to 40 inclusive, what is the sum of the integers from 41 to 80 inclusive, in terms of x?

12 $x/y = k$ and $z/y = 3/k$. What is x/z?

(A) $1/3k$
(B) $3k$
(C) $1/3$
(D) $k/3$
(E) $k^2/3$

Numbers

2.1 :: Integers - Positive/0/Negative

try it yourself

Try this sample question within 30 seconds.

Q1 If a and b are negative integers, which of the following must be true?

I. $a + b$ is a negative integer
II. ab is a positive integer
III. $a/b > a$

(A) none (B) I only (C) II only (D) I and II only (E) I, II, and III

general rules

☑ **Integers** are positive, 0, and negative whole numbers and do not include fractions or decimals.
Examples of integers are -5, -1, 0, 5, and 8.

☑ **Multiplicative Properties of Integers**

(Neg)(Neg) = (Pos)	(-)(-) = (+)
(Neg)(Pos) = (Neg)	(-)(+) = (-)
(Pos)(Pos)(Neg) = (Neg)	(+)(+)(-) = (-)

approach to sample questions

Since a and b are both negative integers,

I: $a + b$ must be a negative integer (Ex: $(-3) + (-4) = -3 - 4 = -7$)
II: ab and a/b are both positive [$(-)(-) = (+)$ and $(-)/(-) = (+)$]
III: ab is positive and is greater than any negative number such as a (Ex: $(-3)(-4) = 12 > -3$)

Therefore, statements I, II, and III are all correct and the answer is (E).

Practice Questions 2.1

1 If neither a nor b is an integer, which of the following could be an integer?

I. $a - b$
II. a/b
III. $\sqrt{a - b}$

(A) None
(B) I only
C) II only
(D) I and II
(E) I, II, and III

2 If s is not an integer, which of the following cannot be an integer?

(A) s/3 (B) 9s/5 (C) 4/s (D) 2s (E) $\sqrt{5s + 2}$

3 If x(m - n) is a negative number, which of the following must be true?

I. x is positive
II. x is positive and $n > m$
III. x is negative and $m > n$

(A) I only
(B) I and II only
(C) II and III only
(D) I and III only
(E) I, II, and III

4 If a and b are negative integers and $a < b$, to which of the following can a/b be equal?

(A) -2 (B) -1/2 (C) 0 (D) 1/2 (E) 2

5 If s, 1/s, t, and 1/t are integers and s and t are not the same integers, which of the following could be a value of st?

(A) 0 (B) 1 (C)2 (D) -1 (E) -2

2.2 :: Integers - Even & Odd / Multiples & Divisibility

try it yourself Try these two sample questions within 30 seconds.

Q1 If K is an odd integer, which of the following is an odd integer?

(A) $K^3 - 3$
(B) $K^2 + 1$
(C) $(K + 1)^2$
(D) $K^2 - 1$
(E) $(K + 2)(K - 2)$

Q2 Of the following, which integer is a multiple of 9?

(A) 123,145
(B) 24,351
(C) 45,628
(D) 87,151
(E) 1,210,212

general rules

☑ **Even & Odd Integers**
Even integers are multiples of 2 or 2k.
Odd integers are not divisible by 2 and are expressed as 2k + 1 or 2k - 1.

E + E = E E + O = O E • E = E E • O = E O • O = O

☑ **Multiples, Divisibility, & Power**
If an integer is divisible by 3 or 9, the sum of its digits is also a multiple of 3 or 9.
If an integer is divisible by 5, its ones' digit is either 5 or 0.
7^n and 9^n can have any number as their ones' digits, while 6^n can only have 6 as its ones' digit.

approach to sample questions

A1 Since K is an odd integer, K + 2, K - 2, and all of their products are also odd. The answer is (E).

A2 The sum of all the digits in the number 1,210,212 is 9, which means that the number itself is divisible by 9. Therefore, the answer is (E).

Practice Questions 2.2

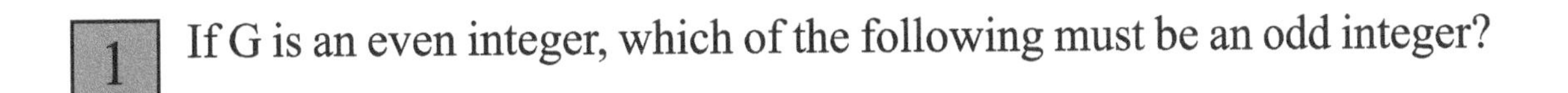

1 If G is an even integer, which of the following must be an odd integer?

(A) 3G + 2
(B) G^2
C) G/2
(D) $(G - 3)^2$
(E) 5G + 2

2 The six digit number 679,3K0 is divisible by 3, 4, 5, and 9 if K is substituted by which of the following digits?

(A) 1
(B) 2
(C) 4
(D) 6
(E) 8

3 When G is an integer, how many odd integers are there in the following list?

G + 2   3G + 1   4G - 1   2(2G - 1)

(A) 0
(B) 1
(C) 2
(D) 3
(E) 4

4 Integers s, t, and u have ones' digits 1, 3, and 9, respectively. Which of the following can be true?

I. s, t, and u can be multiples of 11.
II. s, t, and u can be prime numbers.
III. s, t, and u can be powers of 7.

(A) I only
(B) II only
C) III only
(D) II and III only
(E) I, II, and III

5 If x, y, and z are positive integers and the expression $xy(y - z)$ is an odd number, which of the following numbers can be even?

I. z
II. $x + z$
III. $x + y$

(A) None
(B) I only
(C) II only
(D) I and III
(E) II and III

2.3 :: Prime Factorization

try it yourself Try these two sample questions within 30 seconds.

Q1 Which of the following can equal $6^2 \bullet n$ for some integer n?

(A) $4 \bullet 6 \bullet 10$
(B) $2 \bullet 3 \bullet 15$
(C) $4 \bullet 6 \bullet 16$
(D) $4 \bullet 6 \bullet 8$
(E) $4 \bullet 6 \bullet 12$

Q2 Which of the following has the least number of factors?

(A) 24 (B) 30 (C) 36 (D) 49 (E) 56

general rules

☑ **Factors** of numbers are ones that can evenly divide the number without remainders.
For example, all factors of 12 are 1, 2, 3, 4, 6, and 12

☑ **Prime numbers** are numbers only divisible by 1 and themselves such as
2, 3, 5, 7, 11, 13, 17, 19, 23...
(Remember: 2 is the only even prime number)

☑ **Factor Tree** method to prime factoring

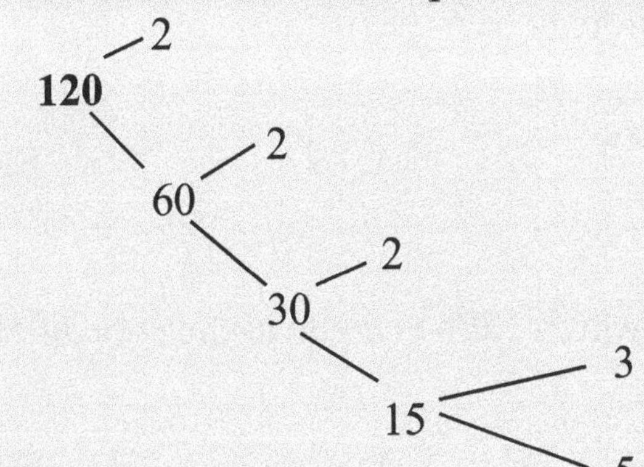

The result of **prime factorization** is
120 = $2 \bullet 2 \bullet 2 \bullet 3 \bullet 5$.

The **prime factors** are 2, 3, and 5.

approach to sample questions

A1 Since 6^2 can be factored out into $2^2 \bullet 3^2$, the answer has to contain at least two 2's and 3's as factors. Therefore, the answer is (E).

A2 49 is 7^2. Therefore, it has only one factor, 7. The answer is (D).

Practice Questions 2.3

1 Which of the following has the least number of prime factors?

(A) 12 (B) 44 (C) 64 (D) 92 (E) 100

2 Which of the following has the most number of even prime factors?

(A) 120 (B) 180 (C) 260 (D) 300
(E) A-D all have the same number of even prime factors

3 How many prime numbers are there between 35 and 55?

(A) 2 (B) 3 (C) 5 (D) 6 (E) 7

4 If $n = 28$, how many distinct prime factors does n^2 have?

(A) 1 (B) 2 (C) 3 (D) 4 (E) 7

5 What is the least positive integer h for which 120h is equal to the cube of an integer?

(A) 120 (B) 180 (C) 225 (D) 240 (E) 300

2.4 :: Fractions & Decimals - L.C.M.

try it yourself

Try these two sample questions within 30 seconds.

Q1 When you compare the values of $\frac{2}{3} - \frac{3}{5}$ and $\frac{4}{5} - \frac{2}{3}$, which is greater?

Q2 If x/3, x/5, and x/4 are integers, which of the following is not necessarily an integer?

(A) x/10 (B) x/15 (C) x/20 (D) x/24 (E) x/30

general rules

☑ When comparing fractions, **multiply them all** by the **least common denominator (L.C.D.)** of all the fractions in comparison, which is the **least common multiple (L.C.M)** between all the denominators.

For example, if you are to compare 1/50 and 2/75, multiply both fractions by the L.C.D., which is the L.C.M. of 50 and 75: 150.
150(1/50) and 150(2/75) become whole numbers: 3 and 4.

☑ If the products of a number and a variable are equal, **divide all sides by the L.C.M.** of all the numbers in equality.

For example, if 2x = 3y = 4z, dividng them by 12 yields x/6 = y/4 = z/3.
Therefore, x > y > z if x, y, and z are positive integers.

approach to sample questions

A1 When comparing fractions, multiply every term by the L.C.D., which is 15 in this case.
10 - 9 = 1 and 12 - 10 = 2. Therefore, the latter value is greater.

A2 In order for x to be divisible by 3, 4, and 5, it has to be a multiple of the L.C.M. of these numbers: 60.
Since 60 cannot be evenly divided by 24, (D) is the answer.

Practice Questions 2.4

1 Which of the following correctly orders 5/3, 7/4, and 3/2 from least to greatest?

(A) 5/3, 7/4, 3/2
(B) 3/2, 7/4, 5/3
(C) 5/3, 3/2, 7/4
(D) 3/2, 5/3, 7/4
(E) 7/4, 3/2, 5/3

2 Of the following, which is the closest in value to 4.20349?

(A) $\frac{4{,}203}{1{,}000}$ (B) $\frac{21{,}017}{5{,}000}$ (C) $\frac{42{,}034}{10{,}000}$ (D) $\frac{21{,}017}{50{,}000}$ (E) $\frac{420{,}347}{100{,}000}$

3 What is the smallest integer that is divisible by 4, 12, and 15 at the same time?

(A) 30 (B) 60 (C) 90 (D) 120 (E) 150

4 If $12x = 30y = 15z$, and x, y, and z are all positive integers, which of the following correctly orders them from least to greatest?

(A) x, y, z (B) z, x, y (C) y, x, z (D) y, z, x (E) z, y, x

5 A combined school with elementary and junior high schools has bells ringing at different intervals for the elementary and junior high school. The junior high bell rings every 60 minutes, while the elementary bell rings every 40 minutes. If the bells ring simultaneously at 8:00 a.m., what is the total number of times the bells will ring simultaneously between 8:00 a.m. and 4:00 p.m. inclusive?

(A) 2 (B) 3 (C) 4 (D) 5 (E) 6

2.5 :: Min/Max Integer Search

try it yourself

Try this sample question within 30 seconds.

Q1 If the product of four different integers is 6, which of the following could be the smallest of these integers?

(A) -6
(B) -3
(C) -1
(D) 2
(E) 3

general rules

☑ The best way to deal with this type of problem is to experiment with different number combinations using a **combination chart.**

☑ The product of several numbers whose sum is constant is maximized **when the numbers are the closest to one another.**

For example, $(3)(5) < (4)(4)$.

approach to sample questions

A1 The following combination chart shows all of the possible multiplicative combinations of different integers whose product yields 6:

-2	-1	1	3
-3	-1	1	2

Therefore, the smallest integer is -3. The answer is (B).

Practice Questions 2.5

1 If x, y, and z are positive integers such that $x < y < z$ and $xyz = 24$, then the least possible value of z is

(A) 2 (B) 4 (C) 6 (D) 8 (E) 12

2 If a and b are integers greater than 1 and $a + b = 15$, what is the smallest possible value of $a - b$?

(A) -13 (B) -11 (C) -1 (D) 1 (E) 3

3 If the product of four different integers is -6, which of the following could be the smallest of these integers?

(A) -6 (B) -3 (C) -2 (D) -1 (E) 1

4 If s and t are positive integers greater than 1 and $s \bullet t = 30$, which of the following must be true?

I. $s + t$ is even
II. Either s or t is a prime number
III. $s + t + 1$ is divisible by 3

(A) I
(B) II
(C) III
(D) I and II
(E) II and III

5 If the average of three different positive integers is 5, what is the greatest possible product of these three integers?

(A) 18 (B) 32 (C) 80 (D) 120 (E) 210

2.6 :: Digit Search

try it yourself Try these two sample questions within 60 seconds.

Q1 In the addition problem below, which of the following cannot be digit Z if X > Y > Z?

(A) 2
(B) 3
(C) 4
(D) 5
(E) 6

$$\begin{array}{r} 1\,X\,4 \\ Y\,2 \\ +\,Z\,7 \\ \hline 3\,0\,3 \end{array}$$

Q2 If x, y, z, and t are different nonzero digits, what is the maximum possible value for t?

(A) 1
(B) 2
(C) 3
(D) 4
(E) 5

$$\begin{array}{r} x \\ y \\ +\quad z \\ \hline 2t \end{array}$$

general rules

☑ The best way to deal with this type of problem is to **experiment with different digit combinations of actual numbers** (see **combination charts** in section 2.5).

☑ It is also useful to remember:
If x is the tens' digit, its actual number is 10x
If x is the hundreds' digit, its actual number is 100x
Etc.

approach to sample questions

A1 According to the problem, $X + Y + Z$ must add up to 19. The different number combinations whose sum yields 19 are listed in the following combination chart:

X	Y	Z
9	8	2
9	7	3
9	6	4
8	7	4
8	6	5

Therefore, the digit Z cannot be 6.
The answer is (E).

A2 Since the maximum possible numbers for x, y, and z are 9, 8, and 7, $9 + 8 + 7 = 24$. Therefore, T can be 4. The answer is (D).

Practice Questions 2.6

1 In the correctly-worked addition of the two-digit numbers below, each letter represents a different positive integer less than 10. Which of the following could represent the number 1SK?

I. 178
II. 134
III. 120

$$\begin{array}{r} XY \\ +\ YY \\ \hline 1SK \end{array}$$

(A) I only
(B) II only
(C) III only
(D) I and II
(E) I, II, and III

2 Which of the following can not be the units digit of 7^x if x is an integer greater than 1?

(A) 1
(B) 3
(C) 5
(D) 7
(E) 9

X, Y, and Z are unique digits in the correctly-worked subtraction problem below. Which decimal is closest to the decimal represented by 0.XYZ?

$$\begin{array}{rr} & 8.740 \\ - & 3.XYZ \\ \hline & 4.XYZ \end{array}$$

(A) 0.620
(B) 0.770
(C) 0.870
(D) 0.920
(E) 0.970

In the addition problem below, which of the following could be the digit C if A and B can be the same number?

I. 1
II. 2
III. 3

$$\begin{array}{rr} & 2A7 \\ & B1 \\ + & C6 \\ \hline & 424 \end{array}$$

(A) I only
(B) II only
(C) III only
(D) II and III
(E) I and III

CHAPTER 2 TEST

1 If x and y are both positive integers, which of the following actions can result in xy?

(A) Multiplying y by 20
(B) Dividing y by 2
(C) Multiplying both x and y by 2
(D) Dividing both x and y by 2
(E) Multiplying x by 2 and dividing y by 2

2 Which of the following could be the product of 65,218 and 384?

(A) 25.043,711
(B) 25,043,712
(C) 25,043,713
(D) 25,043,714
(E) 25,043,715

3 Which of the following has the least number of factors?

(A) 20
(B) 26
(C) 35
(D) 47
(E) 63

4 How many pairs of positive integers (x, y) satisfy the equation $3x + y = 10$?

(A) 1
(B) 2
(C) 3
(D) 4
(E) 5

5 If $p(x)$ is defined as the least prime number greater than x, how many prime numbers are between $p(40)$ and $p(52)$, inclusively?

(A) 0
(B) 1
(C) 2
(D) 3
(E) 4

6 If the average of two different positive integers is 43, then the smallest possible product of the two integers is

(A) 20
(B) 43
(C) 85
(D) 168
(E) 249

7 Let n be the greatest 3-digit integer such that the product of its three digits is 90. What is the units' digit of n?

(A) 2
(B) 3
(C) 4
(D) 5
(E) 6

8 If the degree measures of the interior angles of a triangle are x, y, and z, where $x = y/4$ and $y = 3z/5$, what is the value of z?

(A) 20
(B) 30
(C) 45
(D) 60
(E) 100

9 If x and y are positive integers and $x/4 < y < x/2$, what is the smallest possible value for x?

(A) 3
(B) 4
(C) 5
(D) 6
(E) 7

10 If x and y are integers greater than zero and $x + y = 10$, what is the least possible value of $x - y$?

(A) -12
(B) -11
(C) -10
(D) -9
(E) -8

11 The four-digit number 3,58Q is divisible by 9. What is the value of Q?

(A) 0
(B) 2
(C) 4
(D) 6
(E) 8

Fundamental Rules of Algebraic Operation

3.1 :: Fundamental Rules of Arithmetic & Basic Algebra: Overview

try it yourself Pre-Algebra Diagnostic Test

Q1 $3/4 + 4/5$

Q2 $-2 - (-4) - 3 + 5 + 1 =$

Q3 $2x - x =$

Q4 $3x + 2 - 2(4x - 2) =$

Q5 If $14/x = 35/75$, then $x =$

Q6 If $3/x = 5$, then $x =$

Q7 Solve $2x + 5 = 3x - 7$

Q8 What is 30% of 80?

Q9 If $X^2 = 9$, then $X =$

Q10 $\sqrt{\frac{1}{9} + \frac{1}{16}} =$

fundamental laws of algebra

☑ Add/subtract fractions with different denominators using the **L.C.D.**:
a/b + c/d = (a • d + b • c) / (b • d)

☑ **Minus** -- adding negatives:
-a = + (-a); -(-a) = a; +(-a) = -a

☑ Combining **like-terms**

☑ **Distributive property**

☑ Law of **cross multiplication**: multiply means and extremes together to solve the problem. Use **Rectangular cancellation** to cancel in a rectangular direction before cross-multiplication.

☑ Division on **one side** is multiplication on the **other side** of the equation.

☑ **Equation-solving**

☑ **"%"** means "1/100"
"of" means "multiplication"

☑ Remember **positive and negative roots** of a squared variable.

☑ Be careful when **operating outside a radical expression**.

approach to sample questions

A1 When adding fractions, use the following procedure:

$$\frac{3}{4} + \frac{4}{5} = \frac{3 \bullet 5 + 4 \bullet 4}{4 \bullet 5} = \frac{15 + 16}{20} = \frac{31}{20}$$

A2 -2 - (-4) - 3 + 5 + 1
= -2 + 4 - 3 + 5 + 1 = 5

A3 2x - x = (2 - 1)x = x

A4 3x + 2 - 2(4x - 2)
= 3x + 2 **- 8x + 4** = -5x + 6

A6

$$\frac{\cancel{14}^{\,2}}{x} = \frac{\cancel{35}^{\,\cancel{7}\,1}}{\cancel{75}_{\,15}}$$

x = 30

A6 If $\frac{3}{x} = 5$, then $\frac{3}{5} = x$.

A7 12 = x

A8 30% of 80 = 30 • (1/100) • 80
= 2400/100 = 24

A9 If $x^2 = 9$, x = +3 or -3

A10 $\sqrt{\frac{1}{9} + \frac{1}{16}} = \sqrt{\frac{9 + 16}{144}} = \sqrt{\frac{25}{144}}$

$= \pm \frac{5}{12} \neq \frac{1}{3} + \frac{1}{4}$

some essential short-cuts in algebraic operations

☑ **Rectangular cancelation:** $\frac{x}{24} = \frac{4}{16}$

In the equation above, before multiplying crosswise to solve for an unknown x, we can cancel in the rectangular direction: 24 and 16 or 16 and 4. If we cancel 24 and 16 first, the result becomes $\frac{x}{3} = \frac{4}{2}$, and by canceling 4 and 2 again, we get $\frac{x}{3} = \frac{2}{1}$.

We could also transform the equation to $\frac{x}{4} = \frac{24}{16}$ to solve for x.

☑ **Law of cross multiplication:** $\frac{x}{3} \times \frac{2}{1}$

In the equation above, cross multiplication helps find the answer for x quickly. By cross multiplication, we get $x = 6$.

☑ **Short-cuts in equation solving:**

Linear equation solving

In the equation $5x + 3 = 2x - 6$, gather like terms (x, in this case).
Move the smaller term 2x from the right to the left by changing it into -2x.
Move 3 from the left to the right by changing it to -3.
$5x - 2x = -6 - 3$; $3x = -9$.
The 3 on the left side becomes 1/3 when moved to the right.
$x = 9/(-3) = -3$.

Multiply the other side by its reciprocal: $\frac{2x}{3} = 4$, then $x = 4 \bullet \frac{3}{2} = \frac{12}{2} = 6$

Multiply the other side by the denominator: $\frac{x}{3} = 4$, then $x = 3 \bullet 4 = 12$

Solving a linear equation involving fractions:

Multiply every term by the L.C.D.

Quadratic equation solving:

Factoring method: To solve $x^2 - x = 2$

Move all terms to one side and factor them.
$x^2 - x - 2 = 0$. $(x - 2)(x + 1) = 0$.
Either $(x - 2) = 0$ or $(x + 1) = 0$. Therefore, $x = 2$ or $x = -1$.

Quadratic formula

The quadratic formula is not needed for most problems of the SAT.
However, the formula is $x = \frac{-b \pm \sqrt{b^2 - 4ac}}{2a}$

Practice Questions 3.1

1 If $(x - s)(x + t) = 0$, then $x =$

(A) s or t (B) -s or t (C) s or -t (D) -s or -t (E) s or -s

2 What percent of 240 is 8?

3 If $6/x = y/14$, what is the value of x/y?

(A) 1/84
(B) 3/7
(C) 7/3
(D) 84
(E) Cannot be determined

4 If $4x = x^3$, x can be which of the following?

I. 2
II. 0
III. -2

(A) I only
(B) II only
(C) III only
(D) I and II
(E) I, II, and III

5 If one of the solutions of the equation $x^2 - 2x - c$ is -1, what is the other solution?

(A) -3 (B) 0 (C) 1 (D) 2 (E) 3

3.2 :: Polynomial Product & Factoring

try it yourself Try these two sample questions within 30 seconds.

Q1 If s and t are constants and $x^2 + sx + 5$ factors into $(x + 1)(x + t)$, what is the value of s?

(A) 0 (B) 1 (C) 5 (D) 6 (E) Cannot be determined

Q2 If $(x - y)^2 = x^2 + y^2$, which of the following statements must be true?

I. $y = 0$ II. $(x + y)^2 = (x - y)^2$ III. $xy = 0$

(A) None (B) I only (C) III only (D) II and III (E) I, II, and III

general rules

☑ **Polynomial product expansion rules:**

$(\text{First} \pm \text{Last})^2 = \text{First}^2 \pm 2(\text{First})(\text{Last}) + \text{Last}^2$

$(A + B)^2 = (A + B)(A + B) = A^2 + 2AB + B^2$

$(A - B)^2 = (A - B)(A - B) = A^2 - 2AB + B^2$; thus $(A + B)^2 = (A - B)^2 + 4AB$

$(A + B)(A - B) = A^2 - B^2$

$(A + B)(C + D) = AC + BC + AD + BD$

$(X + A)(X + B) = X^2 + (A + B)X + AB$

☑ **Factoring rules:**

Simple monomial factoring $ax^2 - ax = ax(x - 1)$

Binomial factoring $A^2 - B^2 = (A + B)(A - B)$

Trinomial factoring $A^2 - 2AB + B^2 = (A + B)(A - B)$

Do monomial factoring before the other factoring!

☑ **Quadratic equation solving:** Factoring and the Quadratic Formula

If $x^2 + 2x - 3 = 0$, then $(x + 3)(x - 1) = 0$

Either $x + 3 = 0$ or $x - 1 = 0$; $x = -3$ or $x = 1$

If $x^2 + 2x = 0$, then $x(x + 2) = 0$

$x + 2 = 0$ or $x = 0$; $x = -2$ or $x = 0$.

approach to sample questions

A1 Since $x^2 + sx + 5$ is supposed to be the same as $(x + 1)(x + t)$, we can expand the latter according to the rules of polynomial product expansion

$(x + 1)(x + t) = x^2 + (1 + t)x + t$

Therefore, $t = 5$ and $s = 1 + 5 = 1 + 5 = 6$.

The answer is (D).

A2 From the rule of polynomial product expansion, $(x - y)^2 = x^2 - 2xy + y^2$. But if the problem gives that $(x - y)^2 = x^2 + y^2$, the statement implies that $-2xy = 0$ or $xy = 0$. Furthermore, $(x + y)^2$ is equivalent to $(x - y)^2$ because $2xy = 0$ and $xy = 0$. $(x + y)^2 = x^2 + 2xy + y^2$. Therefore, II and III are both correct; the answer is (D).

Practice Questions 3.2

1 If $25^2 - 4(25)^2 + 3 = (25 - 3)(25 - h)$, then h =

(A) -2
(B) -1
(C) 0
(D) 1
(E) 2

2 If the average (arithmetic mean) of x^2 and y^2 is 47, and $xy = 3$, what is the average of the positive values of x and y?

(A) 1.5
(B) 3
(C) 5
(D) 10
(E) 15

3 If $(x - h)^2 = x^2 + 10x + 25$, what is $(h - 1)^2 - (h + 1)^2$?

(A) 4
(B) 9
(C) 16
(D) 20
(E) 36

If $t > 0$, and s is the only value of x for which $x^2 + tx + 100 = 0$, then $s + t =$

(A) -20
(B) -10
(C) 0
(D) 10
(E) 20

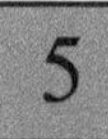

The equation $(x - y)^2 = x^2 - y^2$ is true

(A) for all values of x and y
(B) only for $x = 0$ or $y = 0$
(C) only for $y = 0$ and $x = 1$
(D) only for $y = 0$ and $x = y$
(E) for no values of x and y

3.3 :: Finding a Variable in Terms of Others

try it yourself Try these two sample questions within 30 seconds.

Q1 If $x = 5y$ and $x = 3z + 7y$, what is y in terms of z?

(A) 3z
(B) 2z/3
(C) -2z
(D) 3z/2
(E) -3z/2

TIP: Use substitution to eliminate the variables not needed for the answer.

Q2 If $\frac{t^2 - 9}{(t - 3)^2} = s$ and $s \leq 3$, what is the value of t in terms of s?

(A) (3 + s)/(3 - s)
(B) 9/(s - 1)
(C) (s + 3)/(b - 3)
(D) 3(1 + s)/(1 - s)
(E) (3 + 3s)/(s - 1)

general rules

Remember the following procedure:

- ☑ **Simplify** the original expression as much as possible.
- ☑ **Rearrange** or reshuffle the entire expression.
- ☑ **Collect the like terms** in the variable of interest.
- ☑ **Factor out** in terms of the variable of interest.
- ☑ **Solve** for the variable in terms of the others.

approach to sample questions

A1 Substituting 5y for x in the second expression, ty = 3z + 7y. Therefore, -2y = 3z and y = -3z/2. The answer is (E).

A2 Simplify the expression by factoring and canceling:

$$\frac{(t-3)(t+3)}{(t-3)(t-3)} = s$$

Rearrange the expression.

$$t + 3 = ts - 3s$$

Collect the like terms in the variable of interest, t.

$$t - ts = -3 - 3s$$

Factor in terms of t.

$$t(1 - s) = -3 - 3s$$

Solve for t by the division property for equality.

$$t = (-3 - 3s) / (1 - s)$$

Transform according to the algebraic rule a - b = (-b - a) (see Section 1.5)

$$t = (3 + 3s) / (s - 1)$$

The answer is (E).

Practice Questions 3.3

1 If (1/4)y = -2x + 3z and 4z = 3y, what is x in terms of y?

(A) 3y
(B) 2y/3
(C) -2y
(D) y
(E) -y

2 If $a = 4b = 3c$, then what is the average (arithmetic mean) of a, b, and c in terms of a?

(A) 3a/7
(B) 7a/12
(C) 19a/36
(D) 11a/24
(E) 23a/48

3 $x = 3y + z = 3v$ and $y = z/2$. What is z in terms of v?

(A) 3v/7
(B) 6v/5
(C) 8v/5
(D) 5v/3
(E) 5v/2

4 If $y = \frac{2x + 3z}{4z}$, what is z in terms of x and y?

5 If $\frac{x^2 - 3x + 2}{(x - 1)^2} = y$ and $x \neq 1$, what is the value of x in terms of y?

3.4 :: Operations in Inequality

try it yourself Try these two sample questions within 30 seconds.

Q1 If $3 \leq x < 7$, what is the maximum possible value for $1 - 2x$?

> **TIP**: Most of the maximum/minimum problems in the inequality setting can be handled by working with the bordering numbers, which are 3 and 7 in this case.

Q2 If $6 < su < 8$ and $-8 < ut < -6$, which of the following is true about s and t?

(A) $s > t$
(B) $s < t$
(C) $s = t$
(D) Cannot be determined

general rules

Remember the following **inequality operational rules**:
(All expressions must be positive in order for these rules to apply)

☑ If $x > y$, then $x - y > 0$. (Do not assume $x > 0$ or $y < 0$)

☑ If $3 < -x - 3 < 5$, then $3 + 3 < -x - 3 + 3 < 5 + 3$.
(**Addition/subtraction** property)

☑ If $6 < -x < 8$, then the order is switched and therefore, $-8 < x < -6$.
(**Multiplication/division** property involving negatives requires switching the direction of the inequalities.)

☑ Addition and multiplication of the positives:

$$\begin{array}{rc} & 2 < x < 4 \\ + & 3 < y < 6 \\ \hline & 5 < x + y < 10 \end{array} \qquad \begin{array}{rc} & 2 < x < 4 \\ \times & 3 < y < 6 \\ \hline & 6 < xy < 24 \end{array}$$

☑ Subtraction and division of the positives (**crossing rule**):

$$\begin{array}{r} 2 < x < 4 \\ - \quad 3 < y < 6 \\ \hline 2 - 6 < x - y < 4 - 3 \end{array} \qquad \begin{array}{r} 2 < x < 4 \\ \div \quad 3 < y < 6 \\ \hline 2/6 < x/y < 4/3 \end{array}$$

☑ If x is in the positive range, i.e., $2 < x < 4$, then $1/2 > 1/x > 1/4$.

☑ If $0 < x < 1$, then $x^3 < x^2 < x < \sqrt{x} < 1/x$.
For example, $(1/2)^3 < (1/2)^2 < 1/2 < \sqrt{1/2} < 1/(1/2)$

☑ Solving **quadratic inequality**:

If $x^2 < 2x$, then $x^2 - 2x < 0$.

$$x(x - 2) < 0$$
$$0 < x < 2$$

If $x^2 > 2x$, then $x^2 - 2x > 0$.

$$x(x - 2) > 0$$
$$x < 0 \text{ or } 2 < x$$

Be careful when you multiply mixed inequalities with positives and negatives. The best strategy is to work with the actual numbers to determine the range. For example, if $2 < x < 3$ and $-3 < y < 6$, then the max/min product of xy is $-9 < xy < 18$, not $-6 < xy < 18$.

approach to sample questions

A1 Plug the bordering numbers, 3 and 7, into $1 - 2x$.
Plugging in 3, the answer is -5. Plugging in 7, the answer is -13.
Thus, the maximum is -5 when $x = 3$.
Another method:

$3 \le x < 7$. Multiply all terms by 2, which yields
$6 \le 2x < 14$. Subtracting 1 from all three sides yields
$5 \le 2x - 1 < 13$. Multiply every term by (-1) and the result is
$-13 < 1 - 2x \le -5$. Therefore, the maximum value is -5.

A2 First, in order to use inequality operational rules, every expression must be changed into a positive expression.

$$\begin{array}{r} 6 < su < 8 \\ \div \quad 6 < -ut < 8 \\ \hline 6/8 < -s/t < 8/6 \end{array}$$

(now these are all in positive terms)
$-s/t$ is positive.

However, we cannot determine which value, s or t, is greater or positive.
Therefore, the answer is (D).

Practice Questions 3.4

1 If s - t is negative, which of the following can be true for s and t?

I. $t < s < 0$ II. $s < 0 < t$ III. $s > t > 0$

(A) I only
(B) II only
(C) III only
(D) I and III
(E) I, II, and III

2 If $-1 < x < 0$, which of the following has the smallest value?

(A) $1/x^3$
(B) $1/x^2$
(C) x
(D) x^2
(E) x^3

3 If $x < x^3 < x^2$, then

(A) $x < -1$
(B) $-1 < x < 0$
(C) $0 < x < 1$
(D) $1 < x$
(E) $-1 < x < 1$

4 If k ranges in value from 0.001 to 0.1 and h ranges in value from 0.01 to 1, what is the minimum value of k/h?

(A) 0.001
(B) 0.01
(C) 0.1
(D) 1
(E) 10

If $4 < a < 6$ and $7 < b < 9$, then $1/(b - a)$ is between

(A) 1/3 and 1/5
(B) 1/2 and 1/5
(C) 1/5 and 1/3
(D) 1/2 and 1
(E) 1/5 and 1

What is the greatest integer value for x satisfying $6 < \frac{(42 - 3x)}{5}$?

(A) 1
(B) 2
(C) 3
(D) 4
(E) 5

If $4 < a < 6$ and $-2 < b < 2$, then ab is between

(A) -8 and 12
(B) -8 and 8
(C) -12 and 8
(D) -12 and 12
(E) -2 and 12

3.5 :: Exponential Operations

try it yourself Try these two sample questions within 30 seconds.

Q1 If n is a positive integer, then $(9^n)^2 =$

(A) 3^{4n}
(B) 3^{2n}
(C) 9^{n+2}
(D) 3^{2n}
(E) $81n^2$

> **TIP**: Whenever a non-prime number has a power on it (e.g., 16^2), change the number into a product of prime numbers. For example, $16^2 = (2^4)^2 = 2^8$.

Q2 $5^n + 5^n + 5^n + 5^n + 5^n =$

(A) 5^{5n}
(B) 25^n
(C) 5^{n+5}
(D) 5^{n+1}
(E) $(5^n)^5$

general rules

☑ Remember these **exponential operation rules**:

$x^a \bullet x^b = x^{a+b}$
$x^a \div x^b = x^{a-b}$
$(x^a)^b = x^{ab}$

For example, $(-2x^2y^3)^3 = (-2)^3(x^2)^3(y^3)^3 = -8x^6y^9$

☑ Special exponents

$x^0 = 1$ $\quad$ $x^{-2} = 1/x^2$ $\quad$ $x^{1/2} = \sqrt[2]{x} = \sqrt{x}$ $\quad$ $x^{-1/3} = 1/\sqrt[3]{x}$

Treat numbers with exponents such as 2^x as variables in operations.
For example, $2^x + 2^x = 2(2^x) = 2^{x+1}$.

approach to sample questions

A1 Since 9 is a non-prime number, it needs to be changed into 3^2.
Therefore, $(9^n)^2 = (3^{2n})^2 = 3^{4n}$
The answer is (A).

A2 Treating 5^n as a variable, the sum is $5(5^n) = 5^{n+1}$. The answer is (D).

Practice Questions 3.5

1 Which of the following statements is true?

(A) $(-2s^2t^3)^4 < 0$
(B) $(-2s^2t^3)^4 < (2s^2t^3)^4$
(C) $(-2s^2t^3)^4 = (2s^2t^3)^4$
(D) $(-2s^2t^3)^4 > (2s^2t^3)^4$
(E) None of the above

2 $(0.25)/(0.25)^2 =$

(A) 0.25
(B) 0.4
(C) 4
(D) 40
(E) 400

3 If k is a positive integer, then $(8^k)^2 =$

(A) 2^{8k}
(B) 8^{k+2}
(C) 4^{4k}
(D) 2^{6k}
(E) 16^k

4 $3^{15} - (3^{14} + 3^{13}) =$

(A) $3^{13}(5)$
(B) $3^{13}(14)$
(C) 3^{27}
(D) 6^{13}
(E) 9^{27}

5 If $4^y + 4^y + 4^y + 4^y = 4^{y+k}$, what is the value of k?

(A) 1/4
(B) 1
(C) 4
(D) 16
(E) 4^y

6 If $n > 0$, $\dfrac{9^{n+1}}{9^{n+2}} - \dfrac{9^n}{9^{n+1}} =$

(A) 0
(B) $1/9^n$
(C) 1/9
(D) 1
(E) 9^n

3.6 :: Roots & Radical Operations

try it yourself Try these two sample questions within 60 seconds.

Q1 Simplify $\sqrt{\frac{1}{9} - \frac{1}{16}}$

Q2 s and t are negative. What can $\sqrt{s^2 + t^2}$ not be?

(A) 0 (B) Negative (C) Positive (D) Both A and B (E) None of the above

general rules

☑ Remember the **roots and radical operational rules.**
$\sqrt{4} = 2$ $\sqrt{81} = 9$ $\sqrt[3]{8} = 2$
Caution: if $x^2 = 4$, then $x = +2$ or -2

☑ Squared radical expressions always yield positive results.
If $(\sqrt{x + y})^2 = -x - y$, then $x + y$ must be negative.

☑ When comparing two positive roots, square both of them for comparison.
When comparing 2 and $\sqrt{3}$ by squaring, 4 is greater than 3.
$\sqrt{a^2 + b^2}$ is not the same as $a + b$ and $\sqrt{a/2}$ is not the same as $\sqrt{a}/2$.

☑ When **adding roots**, treat them as variables. For example, $\sqrt{2} + \sqrt{2} = 2\sqrt{2}$

☑ **Rationalization** means removing the radical expression from the denominator of a fraction. For a single term denominator, multiply the same. For a two term denominator, multiply its conjugate.
For example, $\frac{1}{\sqrt{2}} = \frac{1}{\sqrt{2}} \bullet \frac{\sqrt{2}}{\sqrt{2}} = \frac{\sqrt{2}}{2}$ and $\frac{1}{\sqrt{3} - 1} = \frac{1(\sqrt{3} + 1)}{(\sqrt{3} - 1)(\sqrt{3} + 1)} = \frac{\sqrt{3} + 1}{2}$

☑ To solve a **radical equation**, take the following steps:
Move non-radical expressions to other side of the equation.
Square both sides and solve for the unknown.
Check the answer by plugging it back into the equation.
There may be a bogus solution which tags along when the equation is squared.

approach to sample questions

A1 Fractions cannot be removed separately from the radical expression. Therefore, combine it inside the radical expression.

$$\sqrt{\frac{16-9}{144}} = \sqrt{\frac{7}{144}} = \frac{\sqrt{7}}{12}$$

A2 Remember that if $x^2 = 4$, then $x = \pm 2$. However, the square root of 4 is only positive 2. For this reason, $\sqrt{s^2 + t^2}$ can only be positive. Furthermore, in no instance can $\sqrt{s^2 + 2t^2}$ equal 0.
Therefore, the answer is (D).

Practice Questions 3.6

1 If $h = \sqrt{5}$ and $hk = 5$, then what is k?

2 $\sqrt{\frac{1}{25} - \frac{1}{36}} =$

3 If $x^2 = a^2$, what cannot be true about x and a?

(A) $xa < 0$
(B) $x = -a$
(C) $x = a$
(D) $x + a < 0$
(E) All of the statements are true

4 If $1 - \sqrt{2x - 1} = -2$, then $x =$

5 If $0 < a$ and $0 < b$, what is true about $\sqrt{a + b} - (\sqrt{a} + \sqrt{b})$?

(A) Negative
(B) 0
(C) Positive
(D) Cannot be determined

6 If $a > b$, which of the following must be true?

(A) $a + b > 0$
(B) $\sqrt{(b - a)^2} = b - a$
(C) $\sqrt{(a - b)^2} = a - b$
(D) $\sqrt{(a + b)^2} = a + b$
(E) All of the statements are true

3.7 :: Functions / Linear Functions

try it yourself

Try these two sample questions within 30 seconds.

Q1 How many solutions do the equations $x = 3$ and $y = x + 1$ have in common?

Q2 Which of the four quadrants is not touched by the linear graph $2x + 3y = 6$?

(A) Quadrant I
(B) Quadrant I
(C) Quadrant III
(D) Quadrant IV
(E) The graph touches all quadrants.

general rules

☑ A **function** is a relation defined in such a way that for each value x, there exists only one corresponding y value. On a graph, any vertical line through a function has only one intersecting point.

$f(x, y) = \{(1, 2), (2, 2), (3, 2)\}$ is a function.

Domain is all the possible real values for x, which are 1, 2, and 3.

Range is all the corresponding real values for y, which is 2.

$f(x, y) = \{(1, 2), (1, 3), (1, 4)\}$ is not a function.

An equation or inequality of the line containing the point (a,b) holds true when $x = a$ and $y = b$. For example, the inequality $2x + 1 < y$ holds true with the point (-1, 1).

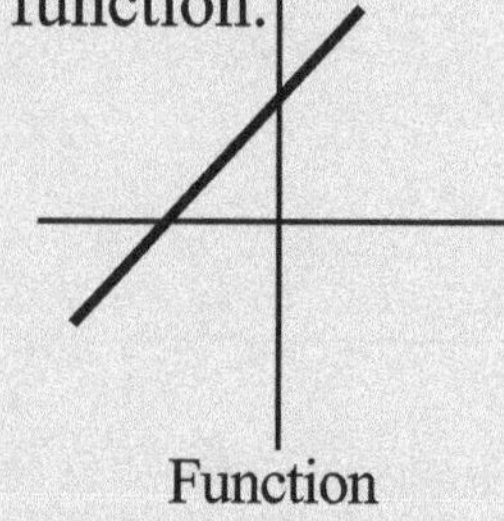
Function

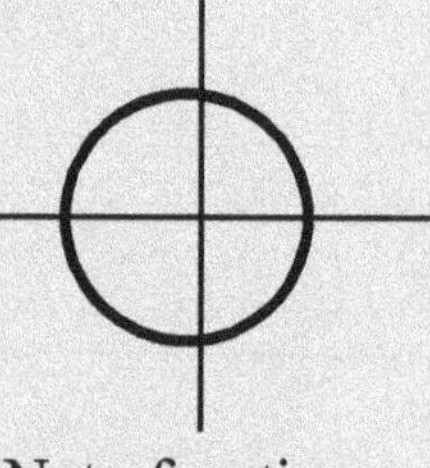
Not a function

When $x = 0$, the corresponding y value is the y-intercept.
When $y = 0$, the coresopnding x value is the x-intercept.

☑ Know how to **graph a linear function**:

standard form: $y = mx + n$, where n = y-intercept

general form: $ax + by = c$

slope (m) $= \frac{a}{b} = \frac{\text{change in y}}{\text{change in x}} = \frac{y_2 - y_1}{x_2 - x_1} = \frac{\text{rise/fall}}{\text{run}} = \frac{5 - 1}{4 - 1} = \frac{4}{3}$

Draw a triangle connecting three points and measure rise/run, starting from the left point (x,y) and moving right.

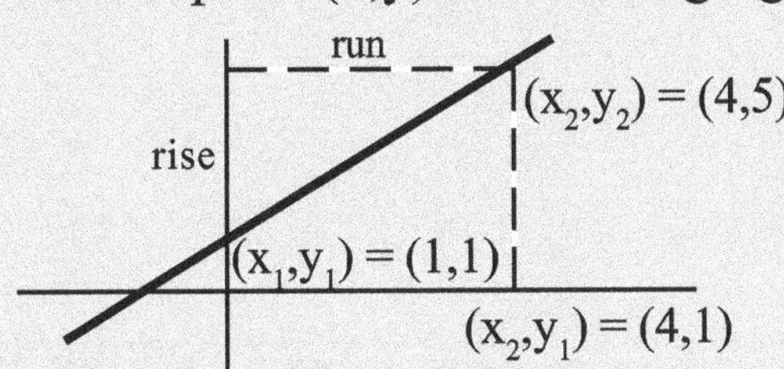

A parallel line has the same slope as the orignal, m, while a perpendicular line's slope is the negative reciprocal, -1/m. In the above example, the slope of a parellel line is 4/3 and the slope of a perpendicular line is -3/4.

Special linear graph: vertical if x = p while y = 0; horizontal if y = q while x = 0.

☑ To find a **solution to a system of equations**, find the point of intersection.

☑ **Quadrants**

(-,+) II	I (+,+)
(-,-) III	IV (+,-)

approach to sample questions

A1 The two given equations have one intersection, and therefore one common solution.

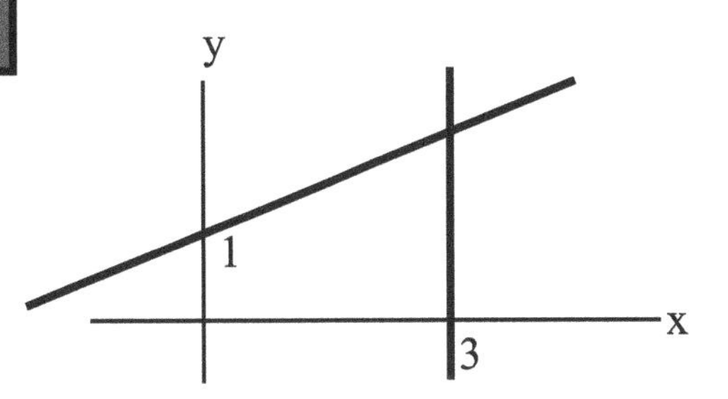

A2 Graphing $2x + 3y = 6$ through its x- and y-intercepts,
When x = 0, the y-intercept is 2.
When y = 0, the x-intercept is 3.
The function does not touch the third quadrant.
The answer is (C).

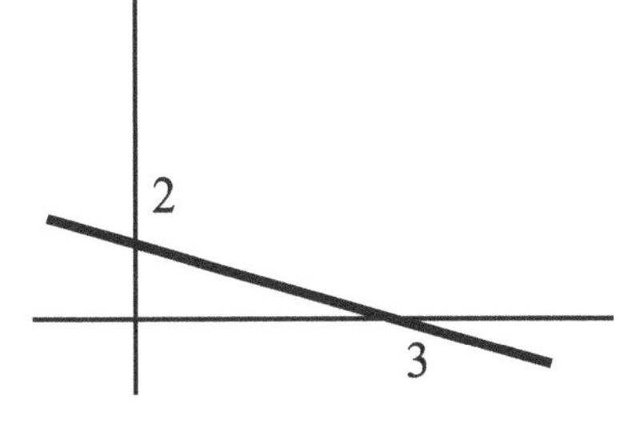

Practice Questions 3.7

1 How many common solutions (x, y) do the equations $2x + 3y = 6$ and $4x + 6y = 6$ have?

(A) None
(B) One
(C) Two
(D) Three
(E) Four

2 If line L is the graph of the equation $3x - 2y = 12$, what is the x-coordinate of the point where the line crosses x-axis?

(A) -6
(B) 0
(C) 3
(D) 4
(E) 5

3 Which quadrant is not touched by the graph of $y = -1$?

(A) I
(B) II
(C) III
(D) IV
(E) I & II

4 How many common solutions do the equations $2x - 5y = -3$ and $10y - 4x = 3$ have?

(A) 0
(B) 1
(C) 2
(D) Infinite
(E) Cannot be determined

5 A function is defined as the relationship where there is only one corresponding y value for each x value.Which of the following relationships below is not a function?

(A)

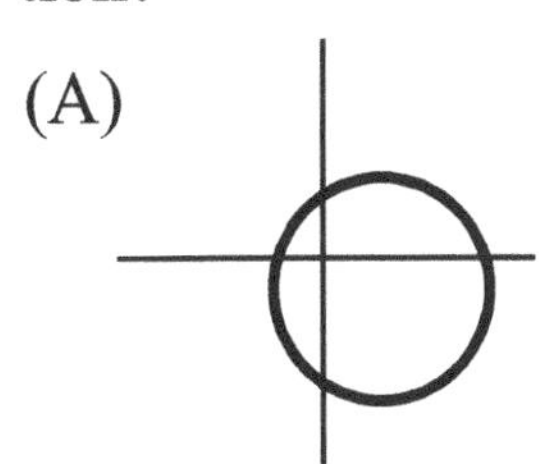

(B)

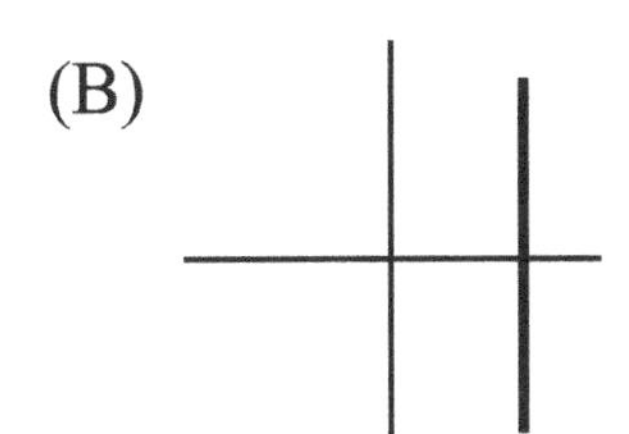

(C)

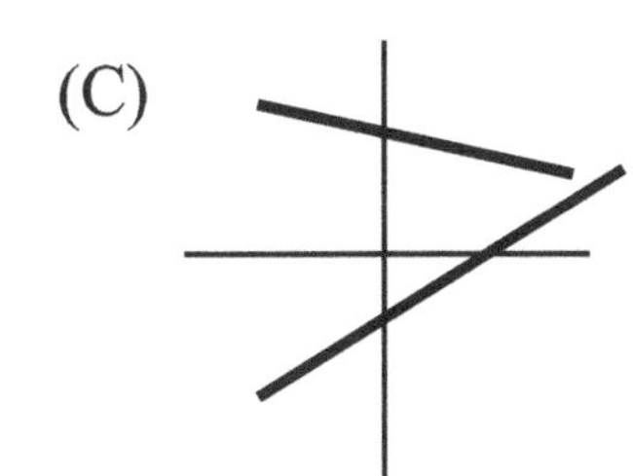

(D)

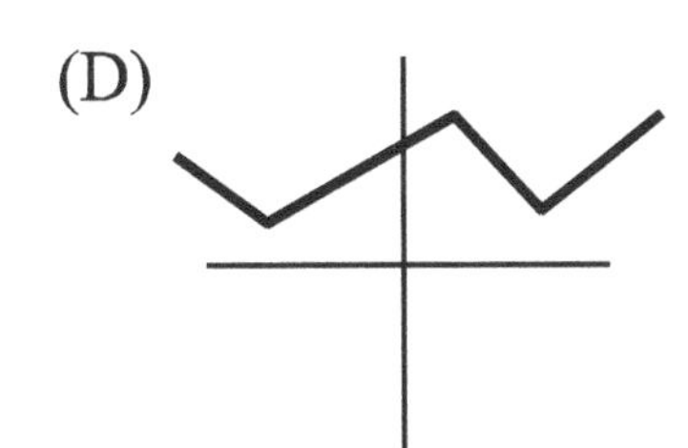

(E)

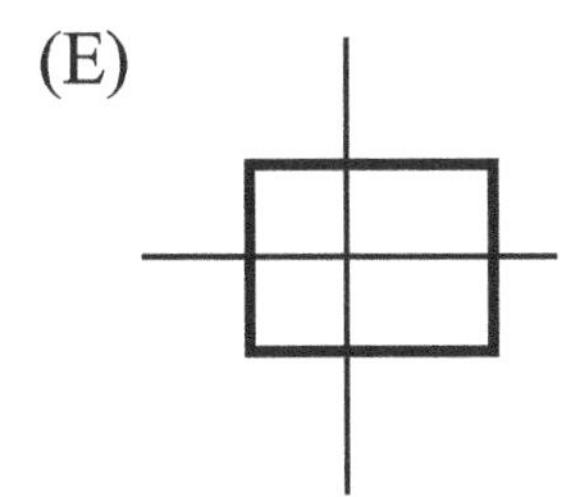

6 In the figure below, the slope of the line through A and B is -4/3. What is the value of t?

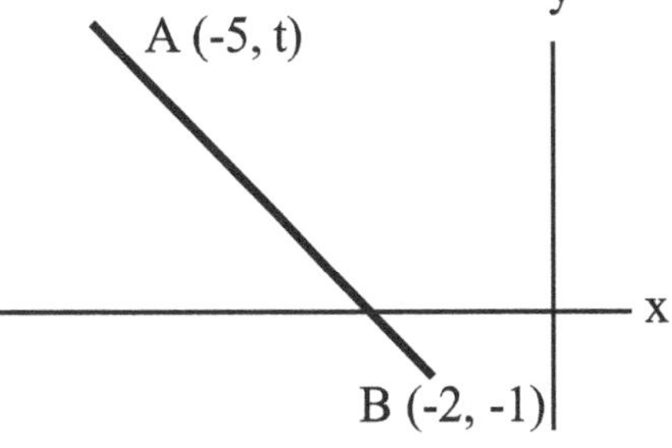

(A) 2
(B) 2.5
(C) 3
(D) 3.5
(E) 4

7 Which of the following statements best represents the graph of the line $y = (-1/2)ax + b$ compared with $y = ax + b$?

(A) It has the same x-intercept, goes in the same direction, and is more steep.
(B) It has a different y-intercept, goes in the opposite direction, and is more steep.
(C) It has the same y-intercept, goes in the same direction, and is less steep.
(D) It has a different x-intercept, goes in the opposite direction, and is less steep.
(E) It has the same y-intercept, goes in the opposite dirtection, and is shifted down.

8 If line T passes through the point (-1, 2) and is perpendicular to the line $x - 2y = 5$, which of the following is true about the x- and y-intercepts of line T?

(A) The x-intercept is greater than the y-intercept.
(B) The x-intercept is less than the y-intercept.
(C) The x-intercept and y-intercept are (5, 0) and (0, -5/2) respectively.
(D) The information provided is not sufficient to determine the x- and y-intercepts.
(E) Both intercepts are same, at the origin.

3.8 :: General Functions & Graphs

try it yourself Try this sample question within 30 seconds.

Q1 If the graph of $f(x) = g(x + 1) - 1$ and $g(x) = 2x + 2$, what is $f(1)$?

general rules

☑ **Finding the value of f(x) when x is known:**
$f(b) = f(x) = y$ when x is plugged in for the value of b.

General Formula: If $f(x)$ = some x, then f(apple) = some apple.
Example: If $f(x) = (x + 10)^3 + x$, then $f(\text{apple}) = (\text{apple} + 10)^3 + \text{apple}$.

Operation of a Function-of-a-Function:
Example: If $f(x) = x + 1$, then $f[g(x)] = g(x) + 1$.

☑ **Finding the intersection(s) of a function and any axis,** or **the intersection(s) of two functions or graphs:**
For a given graph, if the value of x or y is transformed, there are three strategies with which to **draw the new graph**:

Dilate If the size of x's coefficient (m, or slope) increases, draw the graph steeper/narrower.

Shift If x becomes $x - 2$, shift the graph to the right by 2.
If y becomes $y + 3$, shift the graph down by 3.

Reflect If x becomes $-x$, reflect the graph across the y-axis (left/right).
If y becomes $-y$, reflect the graph across the x-axis (up/down).
If y becomes x, rotate the graph around the origin.

Fold If x becomes $|x|$, fold the graph to the right across the y-axis.
If y becomes $|y|$ or $f(x)$ becomes $|f(x)|$, fold the graph upwards across the x-axis.

approach to sample questions

A1 Since f(1) = g(1 + 1) - 1 = g(2) - 1, and g(2) = 2(2) + 2 = 6, f(1) = 6 - 1 = 5.

Practice Questions 3.8

Use the following graph for problems 1-2:

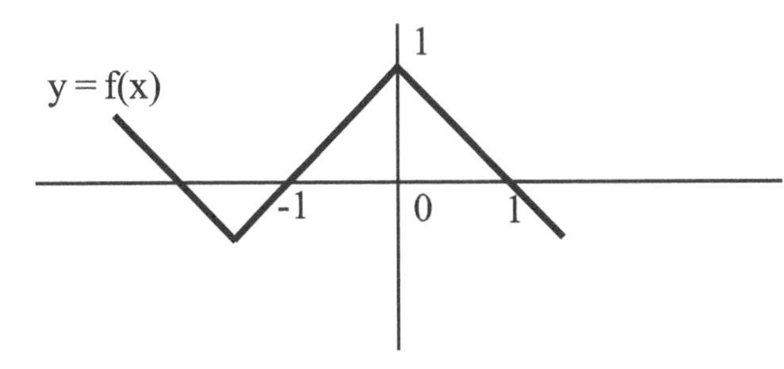

1 If the graph of y = f(x) is shown above and g(x) = -f(x - 1), what is the value of g(2)?

(A) -2
(B) -1
(C) 0
(D) 1
(E) 2

2 In the graph above, if the point p(a, b) is within the area surrounded by f(x) and the x-axis, what is the maximum value of a?

(A) -2
(B) -1
(C) 0
(D) 1
(E) 2

3 If the function $f(x) = x^2 + 1$ and $f(\sqrt{y+1}) = 1$, what is y?

(A) -4
(B) -1
(C) 1
(D) 3
(E) 5

Use the following graph for problems 4-5:

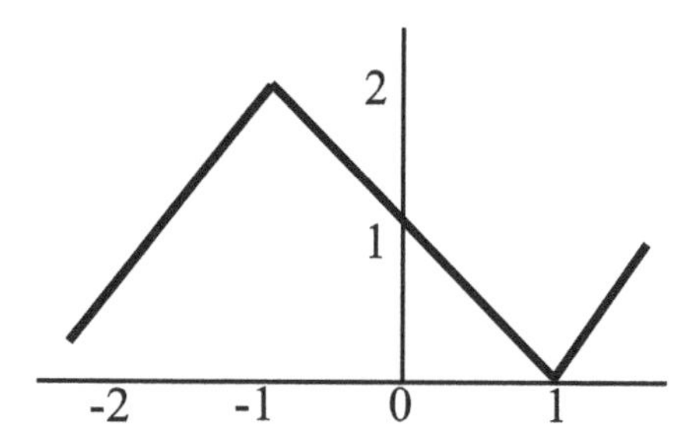

4 If the graph of $y = k(x)$ is shown above and $k(0) = t$, what is the value of $k(t)$?

(A) -1
(B) 0
(C) 1
(D) 2
(E) 3

5 If the graph of $y = k(x)$ is shown above, which of the following could be the graph of $g = k(-x + 1)$?

(A)
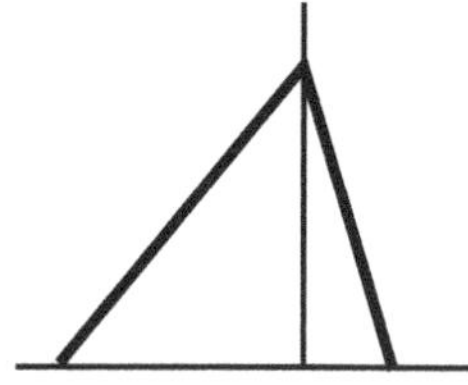

(B)
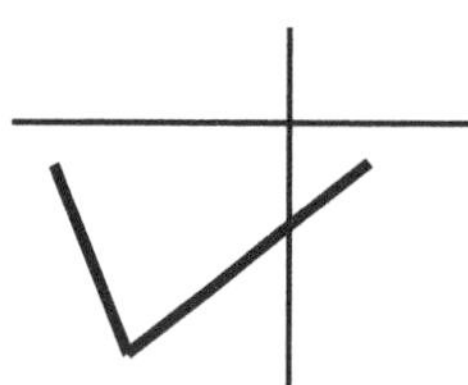

(C)
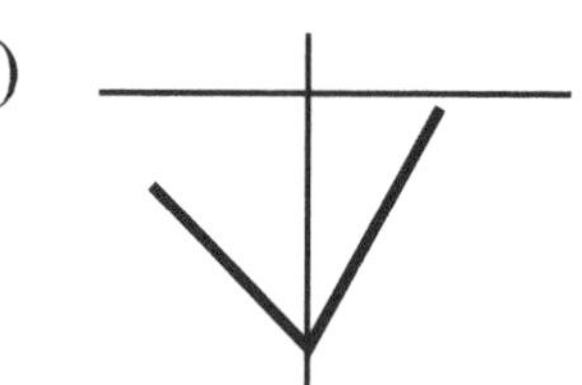

(D)
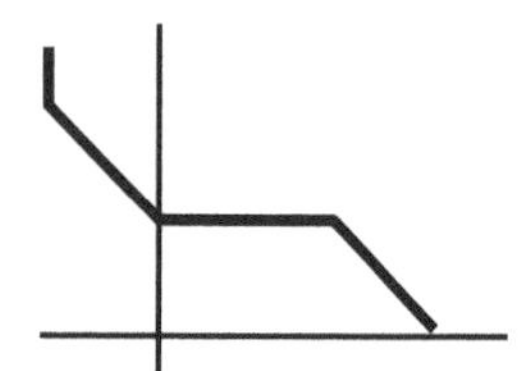

(E)
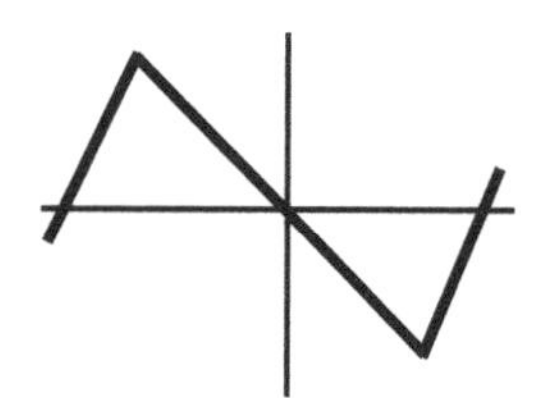

3.9 :: Quadratic / Higher-Order Functions, Graphs, & Eqns

try it yourself Try this sample question within 30 seconds.

Solve $(x + 2)^2 = (x - 3)^2$

general rules

☑ To **graph a quadratic function, transform it:**

If $a > 0$, the parabola opens upwards

If $a < 0$, the parabola opens downwards

General form: $y = ax^2 + bx + c$

to find y-intercept, c

Factored form: $y = a(x - m)(x - n)$

to find x-intercepts, m and n

Complete square form: $y = a(x - h)^2 + k$

to find the vertex (h, k) and the axis of symmetry, $x = h$

☑ To **solve a quadratic equation**, arrange the equation to the factored form (See Section 3.1) and set it equal to zero on the right-hand side:

$(x - m)(x - n) = 0$. Then, $x - m = 0$ and $x - n = 0$.

Therefore, $x = m$ or $x = n$.

Short cut: If $(x - m)^2 = 0$, then $x = m$.

If $(x - m)^2 = (x - n)^2$, then $x - m = \pm(x - n)$.

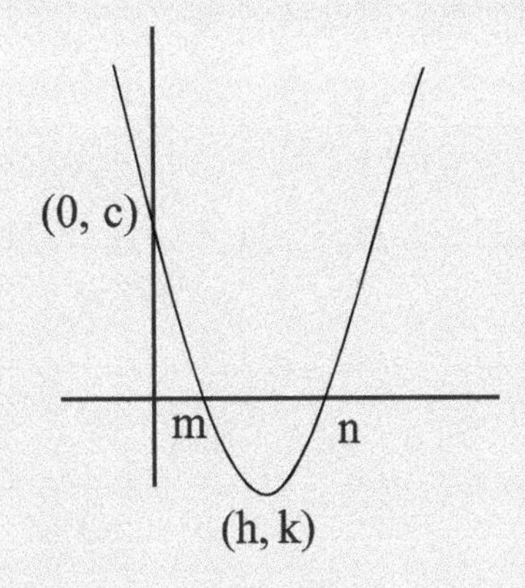

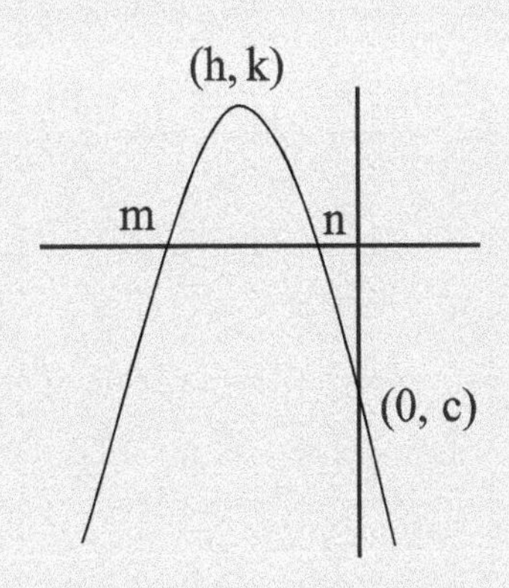

☑ To **find the intersections** of a quadratic function and others, equate the two functions as substitutes for y.

That is, if $y = ax + b$ and $y = cx + d$, equate them: $ax + b = cx + d$.

Most of the quadratic function-related questions in the SAT do not require you to remember the vertex formula. However, you should remember how to read information from the graph of a parabola, such as x-intercepts (when $y = 0$), the y-intercept (when $x = 0$), the axis of symmetry (when $x = h$), and the vertex point among other coordinates.

approach to sample questions

A1 If $(x - h)^2 = (x - k)^2$, then $x - h = \pm(x - k)$. Testing both of these:

$(x + 2) = +(x - 3)$, which can never be true, and

$(x + 2) = -(x - 3)$

So, $2x = 1$ and therefore $x = 1/2$.

Practice Questions 3.9

1 What is/are the value(s) of x satisfying the equation $(x - 1)^2 = (x + 3)^2$?

2 If $x + 3y = 4$ and $x = y^2 + 4$, which of the following is a possible value for y?

(A) -4
(B) -3
(C) 1/3
(D) 1/2
(E) 5

3 If $x + 3$ is a factor of $x^2 - tx - 12$, then $t =$

(A) -4
(B) -1
(C) 1
(D) 3
(E) 5

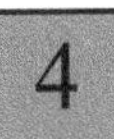

If $x^2 + 5x - 6 = 0$ and $x < 0$, which of the following is not equal to 0?

(A) $x^2 + 4x - 12$
(B) $x^2 - 5x - 6$
(C) $x^2 + 7x + 6$
(D) $x^2 + 4x + 12$
(E) All are equal to 0

If $(x - k)^2 = 9$, what is the sum of the values of the solutions for x?

(A) $k + 9$
(B) $k - 9$
(C) $k + 3$
(D) $k - 3$
(E) $2k$

The table below provides values of quadratic function f for some values x. Which of the following equations could define f(x)?

x	-2	-1	0	1	2
f(x)	-3	0	1	0	-3

(A) $f(x) = x^2 + x - 1$
(B) $f(x) = x^2 - 1$
(C) $f(x) = -x^2 + x + 1$
(D) $f(x) = -x^2 + 1$
(E) $f(x) = x^2 - x + 1$

7 If $f(x) = ax^2 + bx + c$ where $a < 0$, $b > 0$, and $c < 0$, which of the following could be the graph of f?

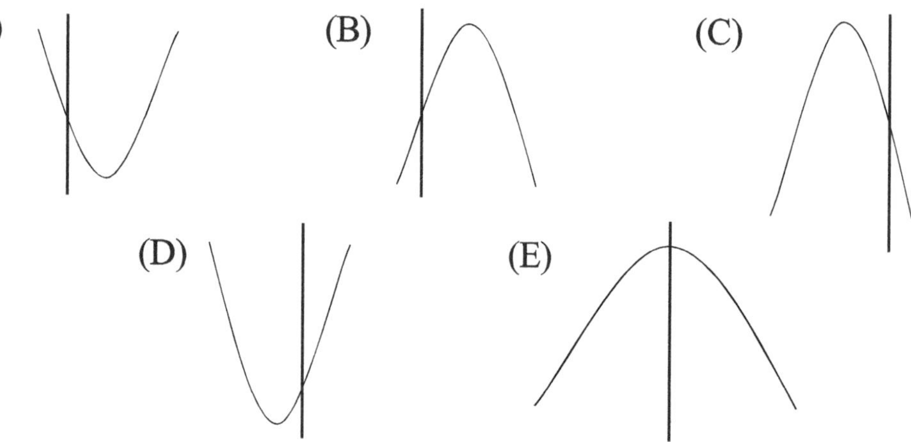

8 If the graph $y = 2x^2 - 2$ intersects line k at (t, 4) and (p, 0), what is the greatest possible value of the slope of k?

9 If (3, t) is one of the points of intersection for the graphs $y = (1/3)x^2 + 2$ and $y = (-1/9)x^2 + k$, where k is a constant, what is the value of k?

10 The graphs $y = 2x^2$ and $y = 4c^2 - 2x^2$, where c is a positive constant, have two intersecting points. If a line segment connecting the two points has a length of 2, what is the value of c?

11 The figure to the right shows the graphs of functions f and g, defined by $f(x) = -x^3 + x$ and $g(x) = f(x + m) + n$, where m and n are constants. What is $m + n$?

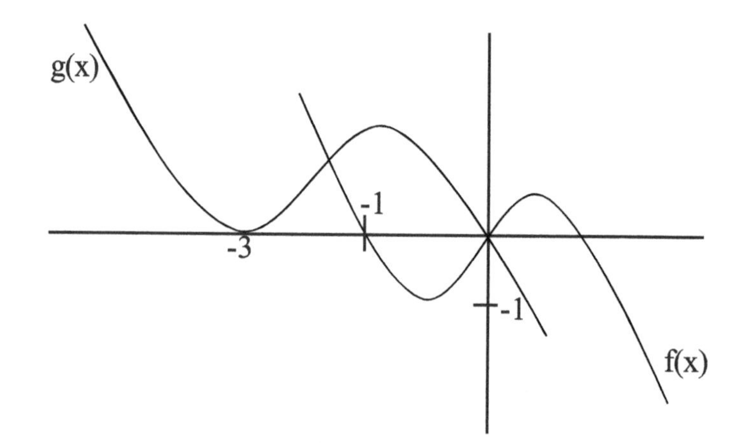

3.10 :: Absolute Value Functions, Equations, & Inequalities

try it yourself Try these two sample questions within 60 seconds.

Q1 What is $|x - 2| + |x|$, if x is greater than 0 but less than 2?

Q2 What is/are the solution(s) for $|x + 2| > 3$?

general rules

☑ The **absolute value** of x is the distance from x to zero on the number line.

For example, $|-5| = 5$

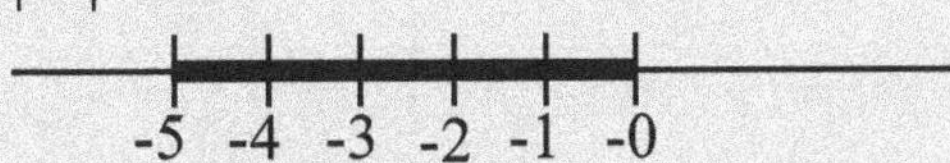

$|x - 2| = 2 - x$ if $(x - 2)$ is negative.

(In other words, if $x - 2 < 0$, $|x - 2| = -x + 2 = 2 - x$)

$|x - k| = |k - x|$. In other words, positive/negative signs and the order of numbers do not matter inside the absolute value.

☑ To **solve absolute value equations and inequalites**, first change any expression to the form of $|A| > k$ or $|A| < k$.

Short cut: If $|A| > k$, then $A > k$ or $A < -k$. If $|A| < k$, then $-k < A < k$.

Orthodox method: Use the piece-wise definition around the value that makes the quantity inside the absolute value equal zero.

Solutions to absolute value inequalities are similar to quadratic inequalites:

If $|A| < x$, then the solutions are between the two values

If $|A| > x$, then the solutions are greater than the larger and less than the smaller values

Example: If $|x - 2| < 3$,

Short-cut: $-3 < x - 2 < 3$, then $-1 < x < 5$.

Orthodox method: If $x - 2 \geq 0$, then $|x - 2| = x - 2$.

Then, $x - 2 < 3$ and $x < 5$. This gives the answer, $2 \leq x < 5$.

If $x - 2 < 0$, then $|x - 2| = -(x - 2)$.

Then, $-x + 2 < 3$ and $x > -1$. This gives the answer, $-1 < x < 2$.

Therefore, combining the two, $-1 < x < 5$.

Example: If $|x - 2| > 3$,

Short-cut: $x - 2 > 3$ or $x - 2 < -3$. Therefore, $x > 5$ or $x < -1$.

Orthodox method: If $x - 2 \geq 0$, then $|x - 2| = x - 2$.

Then, $x - 2 > 3$ and $x > 5$. This gives the answer, $5 < x$.

If $x - 2 < 0$, $|x - 2| = -(x - 2)$.

Then, $-x + 2 > 3$ and $-x > 1$. This gives the answer, $x < -1$.

Therefore, combining the two, $x > 5$ or $x < -1$.

☑ **Graph** absolute value functions piece-wise around the point defined by the function.

Example: If $y = |x - 2|$, graph the function around the point $x = 2$.

$y = x - 2$ when $x \geq 2$ and $y = -x + 2$ when $x < 2$

If $y = -|x + 4|$, graph the function around the point $x = -4$.

If $y = |(x + 2)(x + 4)|$, graph around the points $x = -2$ and $x = -4$.

approach to sample questions

A1 Since $0 < x < 2$, $x - 2 < 0$. Therefore, $|x - 2| = -x + 2$ and $|x| = x$.
Then, $|x - 2| + |x| = -x + 2 + x = 2$.

A2 Short-cut: If $|x + 2| > 3$, then $x + 2 > 3$ or $x + 2 < -3$.
Therefore, $x > 1$ or $x < -5$.

Orthodox method: If $x + 2 > 0$, then $|x + 2| = x + 2$.

Then, $x + 2 > 3$, so $x > 1$.

If $x + 2 < 0$, then $|x + 2| = -(x + 2)$.

Then, $-x - 2 > 3$, so $x < -5$.

Therefore, $x > 1$ or $x < -5$.

Practice Questions 3.10

1 What is the value of a - b in the absolute value equations below when $b < 0 < a$?

$$|a + 5| = 7$$
$$|b - 2| = 3$$

2 Which of the following coordinates satisfy the function $|y| - |x| = 1$?

(A) (2, 1)
(B) (-2, 3)
(C) (3, -2)
(D) (4, 3)
(E) (-5, 4)

3 A survey wants to collect data specifically from students earn a score of T, which is between 650 and 750 in the quantitative section of the SAT. Which of the following inequalities can be used to express these search parameters?

(A) $|T - 650| < 50$
(B) $|T + 700| < 50$
(C) $|T - 750| < 50$
(D) $|T + 700| > 50$
(E) $|T - 700| < 50$

4 If $|-2 - x| < 5$, which of the following is a possible value of x?

(A) -7
(B) -5
(C) 3
(D) 5
(E) 7

5 Which of the following can be a correct piece-wise definition of the function $h(x) = |x + 2|$?

(A) $h(x) = x + 2$ when $x \geq 0$ and $x - 2$ when $x < 0$
(B) $h(x) = x + 2$ when $x \geq 2$ and $-x + 2$ when $x < 2$
(C) $h(x) = x + 2$ when $x \geq -2$ and $-x - 2$ when $x < -2$
(D) $h(x) = x + 2$ when $x \geq -2$ and $x - 2$ when $x < 2$
(E) $h(x) = x + 2$ when $x \geq 2$ and $-x + 2$ when $x < 0$

6 Find the solutions to $|x + 3| > x$ for the value of x, when $-3 < x < -1$.

7 Find the solutions for $|5 - 2x| > 5$.

CHAPTER 3 TEST

1 If $x/y + y/x = 5$, what is the value of $(x + y)(1/x + 1/y)$?

(A) 3
(B) 7
(C) 10
(D) 15
(E) 25

2 If s and t are constants and $x^2 + sx + 8$ factors into $(x + 1)(x + t)$, what is the value of s?

(A) 1
(B) 2
(C) 4
(D) 8
(E) 9

3 If the degree measures of the angles of a triangle are x, y, and z, where $x = y/3$ and $y = 3z/5$, what is the value of z?

(A) 20
(B) 30
(C) 50
(D) 60
(E) 100

4 If $2 < x < 5$ and $7 < y < 10$, then $1/(y - x)$ is between

(A) 1/8 and 1
(B) 1/4 and 1
(C) 1/5 and 2/3
(D) 1/8 and 1/2
(E) 1/2 and 2/3

5 If x and y are positive integers and $x/4 < y < x/2$, what is the smallest possible value for x?

(A) 3
(B) 4
(C) 5
(D) 6
(E) 7

6 If n is a positive integer, then $(4^n)^3$ is NOT equivalent to

(A) 4^{3n}
(B) 2^{6n}
(C) 12^n
(D) 64^n
(E) None of the above

7 How many solutions are there between the equations $x = -1$ and $y = 3$?

(A) 0
(B) 1
(C) 2
(D) 3
(E) Cannot be determined

8 If $a^2 + b^2 = 18$ and $ab = 6$, then $(a - b)^2 =$

(A) -12
(B) -6
(C) 0
(D) 6
(E) 12

9 Which of the following could be the values of x and y for $x^2 + y^2 > 25$?

(A) (-5, 0)
(B) (-3, -4)
(C) (-4, -1)
(D) (3/4, 3)
(E) (4, -4)

10 If function $y(x) = |x + 1|$ and $g(x) = 1$, what are the values of x where the two graphs intersect?

(A) -1, 2
(B) 0, 2
(C) -1, 2
(D) -2, 0
(E) -2, -1

11 The equations $x + y = 10$ and $x + y = 15$ have how many solutions (x, y) in common?

(A) None
(B) Exactly one
(C) Exactly two
(D) Exactly four
(E) More than four

12 If $2 - \sqrt{3x - 5} = -2$, then x =

(A) 1
(B) 5/3
(C) 3
(D) 11/3
(E) 7

13 If the graph of the function $y = 2(x - 1)^2$ is a parabola, what can you say about its shape in relation to the graph of the function $g(x) = -x^2$?

(A) Opens in the same direction, shifted to the right, and with a less steep slope
(B) Opens in the opposite direction, shifted to the right, and with a less steep slope
(C) Opens in the same direction, shifted to the left, and with a steeper slope
(D) Opens in the opposite direction, shifted to the right, and with a steeper slope
(E) Opens in the opposite direction, shifted to the left, and with a less steep slope

14 If $ax + by = ay + bx$ and $a \neq b$, find x in terms of the other variables.

(A) (a - b) / y
(B) y / (a - b)
(C) (a - b) / y
(D) y
(E) 1 / y(a - b)

Special Types of Algebra Problems

4.1 :: Extra-Terrestrial Problem

try it yourself

Try these two sample questions within 30 seconds.

Q1 If I ♡ U = $\frac{I + U}{I - U}$, then what is 5 ♡ 10?

(A) -5
(B) -3
(C) 1
(D) 3
(E) 15

Q2 Let <n = n/2 if n is even and <n = 2n if n is odd.
What is <7 • <20?

(A) <140
(B) <70
(C) <270
(D) <1400
(E) <280

general rules

Don't panic if you see some strange symbols on the test. This kind of math may seem extra-terrestrial, but with our super intelligent brains, we humans can decipher what these symbols mean.

Example: If I ♡ U = IU + I + U,
then 3 ♡ 4 = 12 + 3 + 4 = 19.

Example: If @[x] is defined as the least integer greater than or equal to x,
then @[-3.2] = -3

Example: If [x] is 2 more than the number of digits in the integer x,
then [1000] = 4 + 2 = 6.

approach to sample questions

A1 Following the definition, $5 \heartsuit 10 = \frac{5 + 10}{5 - 10} = \frac{15}{-5} = -3$. The answer is (B).

A2 Since <7 is 14 and <20 equals 10, <7 • <20 = 14 • 10 = 140.
Choose the answer that yields 140.
Since <240 = 140, the answer is (E).

Practice Questions 4.1

1 If the operation @ is defined for all numbers x and y by the equation $x @ y = \frac{x - y}{3}$, then 3 @ (11 @ 2) =

(A) -5
(B) -3
(C) 0
(D) 1
(E) 3

2 For any number x, x^ is defined as the greatest integer that is less than x. What is (3.7)^ + (-3)^ ?

(A) -3
(B) -1
(C) 0
(D) 1
(E) 3

3 Let x*** = $x - x^2$ for all non-negative integers x. Which of the following equals $x^2 - x +$ x*** ?

(A) -3***
(B) -1***
(C) 1***
(D) 2***
(E) 3***

If $x \# y = (x + y)(x - y)$ for all real numbers, then which of the following must be true?

I. $x \# y = y \# x$
II. $x \# 0 = 0 \# x = x^2$
III. $x \# -y = x \# y$

(A) I only
(B) II only
(C) III only
(D) II and III
(E) I, II, and III

5

For all positive integers a and b, the expression a!b is defined as a • b plus the greatest prime factor of a • b. For example, 3!2 = 6 + 3 = 9. If 4!b = 55, then what is b?

(A) 10
(B) 11
(C) 12
(D) 13
(E) 14

6

For all positive integers n greater than 1, let >n be the total number of all the prime numbers from 1 to n. For example, >15 = 6 because there are a total of 6 prime numbers of 15: 2, 3, 5, 7, 11, and 13. What is >97 equivalent to?

(A) >105
(B) (>105) -1
(C) (>105) -2
(D) (>105) -3
(E) (>105) - 4

4.2 :: Division & Remainder Problem

try it yourself Try this sample question within 30 seconds.

Q1 When x is divided by 5, the remainder is 2.
When y is divided by 5, the remainder is 1.
What is the remainder when xy is divided by 5?

general rules

There are three approaches to this type of problem:

☑ **Simple approach**: experimenting with actual numbers

Example: If x is divided by 5, the remainder is 2. What is the remainder when $x + 3$ is divided by 5?

In this case, take the numbers 2, 7, 12, ...
When we add 3 to these numbers, the numbers become 5, 10, 15, ...
When these are divided by 5, the remainder is 0.

☑ **Formal approach**: using the division rule

In the above example, the equation can be as follows:

Since $x \bullet 5 = Q$ with a remainder of 2, $x = 5Q + 2$.
Therefore, $x + 3 = 5Q + 2 + 3 = 5Q + 5$.
Dividing $5Q + 5$ by 5 yields $Q + 1$, and the remainder is 0.

The good thing about this approach is that the quotient can also be determined.

☑ **Remainder approach**:

Working with a remainder yields the same results as those determined via the formal approach, but within a shorter period of time. Adding 3 to the remainder 2, the new remainder becomes $2 + 3 = 5$. Dividing this new remainder of 5 by 5 leaves us with a remainder of 0.

approach to sample questions

A1 Think of actual numbers for x and y. Assign 7, 12, ... to x and 6, 11, ... to y. Therefore, xy = (7)(6) = 42. Divide 42 by 5, and the remainder is 2.

Practice Questions 4.2

1 Chris had s number of baseball cards. He decided to distribute them equally to six of his friends. After he gave each of them x cards, he decided to keep the remainder for his future children. How many cards did Chris keep for his children?

(A) s/6x
(B) 6s/x
(C) s/7x
(D) s - 6x
(E) s - 7x

2 If n is divided by 6, the remainder is 2. If (3n + 1) is divded by 6, what is the remainder?

(A) 1
(B) 2
(C) 3
(D) 4
(E) 5

3 There were between 110 and 120 people in a theatre. When they were counted by 7's when entering the theatre, there were 6 left over. When they were counted by 6's when exiting the theatre, there were 4 left over. How many people were in the theatre?

(A) 111
(B) 113
(C) 114
(D) 116
(E) 118

When k is divided by 6, the remainder is 2.
When h is divided by 6, the remainder is 4.
What is the remainder when k + h is divided by 6?

(A) 0
(B) 1
(C) 2
(D) 4
(E) 5

5

When a number is divided by 6, the remainder is 5. What is the remainder if the number is divided by 3?

(A) 0
(B) 1
(C) 2
(D) 3
(E) Cannot be determined

4.3 :: Sequence & Series / Pattern Search

try it yourself Try these two sample questions within 30 seconds.

Q1 In the sequence x, y, z, 57, when any one of the first three terms is subtracted from the term immediately following it, the result equals x + 3. What is the value of x?

(A) 9 (B) 10 (C) 11 (D) 12 (E) 13

Q2 The sequence 3, -6, 9, -12, ... is formed by taking all positive multiples of 3 in increasing order and changing each even term to its negative. Then what is the sum of the first 100 terms?

general rules

☑ **Simple experimental approach**:
Working with actual numbers in search of a pattern is the most popular approach. However, a complete list of the numbers is not necessary to find the pattern. Usually, the first couple and the last couple of numbers in the sequence will suffice.

☑ **Sequence and series formula**:
This is not required because Algebra II is not required for the test.

Arithmetic sequence

$A_n = a_1 + (n - 1)d$, where A_n is the general term, a_1 is the first term, n is the number of terms, and d is the common difference.

Arithmetic sum (series)

$S_n = \frac{n(a_1 + a_n)}{2}$

Geometric sequence & series

$A_n = a_1 r^{n-1}$

$S_n = a_1(1 - r^n) / (1 - r)$

where r is the common ratio

approach to sample questions

A1 The question states that in the sequence x, y, z, 57, when any one of the first three terms is subtracted from the term immediately following it, the result equals x + 3.

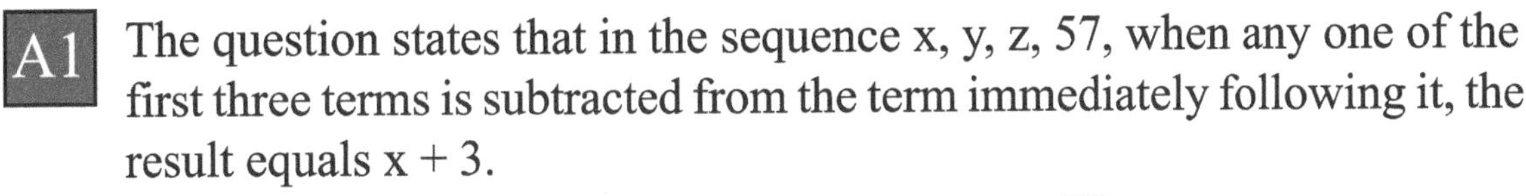

The common difference is x + 3.
From the above relationship, $x + 3 (x + 3) = 57$. $4x + 9 = 57$.
$4x = 48$; $x = 12$. The answer is (D).

A2 Organizing the sequence into pairs (Bird's-Eye View reuniting strategy), we get (3, -6), (9, -12), ... (297, -300).
Therefore, the sum of the first 100 terms is (-3) • (50 pairs) = -150.

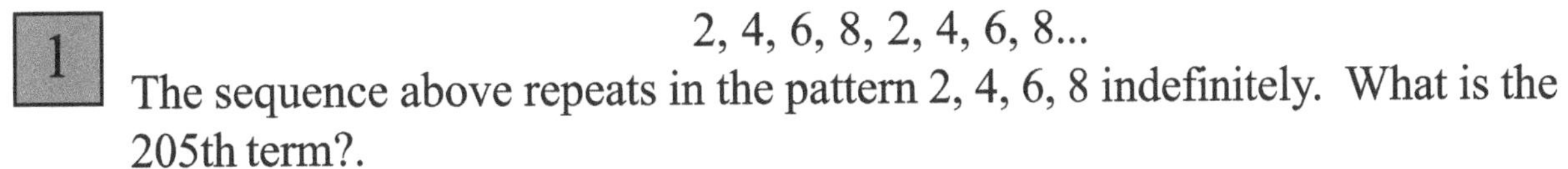

Practice Questions 4.3

1

2, 4, 6, 8, 2, 4, 6, 8...

The sequence above repeats in the pattern 2, 4, 6, 8 indefinitely. What is the 205th term?.

(A) 2
(B) 4
(C) 6
(D) 8

2 The first three numbers on a certain list are 7, 8, and 9. Starting with the fourth number on the list, each number is the sum of the three that immediately precede it. If the nth number on the list is the first to exceed 200, what is the value of n?

(A) 7
(B) 8
(C) 9
(D) 10
(E) 11

3

$$10 + 5 = (5/2)(2)(3)$$
$$15 + 10 + 5 = (5/2)(3)(4)$$
$$20 + 15 + 10 + 5 = (5/2)(4)(5)$$

For each of the three sums shown above, let n represent the number of multiples of 5 being added. Which of the following is an expression for each of the three sums in terms of n?

(A) $2n + 1$
(B) $5n$
(C) $n(n + 1)$
(D) $(5/2)(n)(n + 1)$
(E) $n(n + 1)$

A yo-yo, when dropped, rebounds 2/3 of the distance it falls from its starting height. On each bounce thereafter, it rebounds 2/3 of the previous height. If the yo-yo is dropped from a height of 3 feet, how many vertical feet will it have traveled in total after it swings down the third time?

(A) 22/3
(B) 25/3
(C) 28/3
(D) 29/3
(E) 32/3

3, 9, 27...

In the sequence above, each term after the first is found by multiplying the preceding term by 3. The sum of the first n terms of the sequence is equal to 1/2 times 3 less than the value of the $(n + 1)$th term. What is n when the sum of the first n terms is 1092?

(A) 4
(B) 5
(C) 6
(D) 7
(E) 9

6

The sequence k_1, k_2, k_3, k_4 is such that $k_n = k_{n-1} - 3$ for $1 < n < 5$. If $k_4 = 25$, what is the value of k_1?

(A) 12
(B) 16
(C) 22
(D) 34
(E) 40

4.4 :: Logic/Set/Counting Problem

try it yourself Try these two sample questions within 30 seconds.

Q1 If floors 17 to 100 of a 100-story building are not damaged from a fire, how many floors are not damaged?

(A) 82 (B) 83 (C) 84 (D) 85 (E) 81

Q2 A bag contains 16 black or white tickets that are either with or without holes. If exactly 1/2 have holes, 1/4 are black, and exactly 5 are white with no holes, how many tickets are black with holes?

general rules

Imagine the situation and write it out, drawing a picture if necessary.

☑ **Logic: Remember converse, inverse, and contrapositive**

If the **original** statement is "if I am a human, I am an animal,"

the **converse** would be "if I am an animal, I am a human,"

the **inverse** would be "if I am not a human, I am not an animal,"

the **contrapositive** would be "if I am not an animal, I am not a human."

The original is the same as the contrapositive, while the inverse is the same as the converse.

☑ **Set problem**:

A = { 0, 1, 2, 3, 4, 5, 6, 7 }

B = { x | x is a prime number }

The set of elements in B that belong to A is { 2, 3, 5, 7}.

☑ **Consecutive number counting formula**:

The actual number in a sequence is one more or less than the difference between the two numbers, depending on whether these numbers are included.

\# of sequence = upper limit - lower limit ± 1 (+ if inclusive, - if exclusive)

For example, there are 91 numbers between 10 and 100 inclusively, and 89 numbers exclusively.

approach to sample questions

A1 The case includes the 17th and 100th floors, and therefore, is inclusive. 100 - 17 + 1 = 84. The answer is (C).

A2 As you can see from the diagram below, the number of black tickets with holes is 4 - 3 = 1.

WH 12 - 5 = 7	WN 5	16 - 4 = 12
4 - 3 = 1 **BH**	8 - 5 = 3 BN	(1/4) • 16 = 4
(1/2) • 16 = 8	16 - 8 = 8	

Practice Questions 4.4

1 The statement “If I was born in the USA, I am a citizen of the USA” is true. Which of the following is also true?

(A) If I am a citizen, I was born in America.
(B) If I am not a citizen of the USA, I was not born in America.
(C) If I was not born in America, I am not a citizen of the USA.
(D) All of the above.
(E) None of the above.

2 In a class of 20, there were 32 awards given out during the entire school year. Which of the following must be true?

(A) Each student received at least one award.
(B) One of the students received exactly two awards.
(C) At least one student received more than award.
(D) Twelve students received exactly two awards.
(E) Some students received no awards.

3 In a certain chess game, a single move will advance a player either 1, 3, or 4 spaces. What is the minimum number of moves required to advance 23 spaces?

(A) 4
(B) 5
(C) 6
(D) 7
(E) 8

4 A pencil case contains 20 pencils that are either new or used and are either red or green. If exactly 1/4 of them are new, 1/4 are red, and exactly 12 are used and green, how many are new and green?

(A) 2
(B) 3
(C) 4
(D) 5
(E) 6

5 In a game where there are 10 cards on the table, the player who removes the last card wins. Three players in the order of A, B, and C remove at least 1 but no more than 5 cards at a turn. If Player C removes only 1 at her turn, how many cards must player A remove on her first turn in order to make sure that she wins the game?

(A) 1
(B) 2
(C) 3
(D) 4
(E) 5

6 How many multiples of 3 are there between 3 and 999 (not including 3 or 999)?

(A) 331
(B) 332
(C) 333
(D) 998
(E) 999

4.5 :: Number of Cases / Probability

try it yourself Try these two sample questions within 60 seconds.

Q1 A telephone area code is a three-digit number. If no code can begin with 0 or 1, and 911 and 411 are excluded, how many different area codes are possible?

(A) 26 (B) 98 (C) 510 (D) 798 (E) 998

Q2 Bag P contains 15 marbles, 5 of which are green. Bag Q contains 30 marbles, 15 of which are blue. If one marble is drawn at random from each bag, what is the probability that the marble from bag P is green and that the one from bag Q is blue?

general rules

☑ **Probability** formula

(see **Venn Diagrams** in Section 6.1)

The probability of at least 1 case (Case 1 or 2):

p(either Case 1 or Case 2) = p(Case 1) + p(Case 2) - p(both Case 1 and Case 2)

For example, the probability of earning a perfect score on at least one section (verbal or quantitative) of the exam = p(perfect verbal score) + p(perfect quantitative score) - p(perfect verbal and quantitative scores)

Mutually exclusive or **Disjoint**:

p(both Case 1 and Case 2) = 0; Case 1 cannot occur if Case 2 does, and vice versa.

p(either Case 1 or Case 2) = p(Case 1) + p(Case 2)

For example, if you can only earn a perfect score in either the verbal or quantative sections, but never both.

Simultaneous/independent:

p(Case 1) does not influence p(Case 2) because they occur independently.

P(both Case 1 and Case 2) = p(Case 1) • p(Case 2)

For example, earning a perfect score in the quantitative section is not influenced by how you perform in the verbal section.

☑ **Writing it out** (most common approach)
Although this may be time-consuming, most of the probability problems on the exam are not too complicated to be solved in this way.

☑ **Probability** (maximum = 1) = $\dfrac{\text{\# of cases favored}}{\text{\# total possible cases}}$

☑ **Permutation:** The problem involves **one-time choice and arrangement,** in which the order matters (i.e. phone numbers, lock combinations)

$$n! = n(n-1)(n-2)\ldots(2)(1)$$

$$_nP_r = n(n-1)(n-2)\ldots(n-r+1)$$

Example: selecting and arranging 5 cards from a 52-card deck is $_{52}P_5$.

☑ **Combination:** The problem involves **one-time choice only**, in which the order does not matter (i.e. groups of people)

$$_nC_r = \frac{_nP_r}{r!} = \frac{n!}{r!(n-4)!}$$

Example: selecting 3 officers from 50 people is $_{50}C_3$.

approach to sample questions

A1 Write out all of the possible numbers for the first digit starting from 2 to 9. Only these 8 are possible numbers for the first digit of the area code. There are 10 possible numbers for the second digit and 10 for the third digit. The numbers are selected simultaneously, so multiply them:
8 • 10 • 10 = 800.
Subtracting the 2 cases of 911 and 411, 800 - 2 = 798. The answer is (D).

A2 The probability of choosing green is 1/3, and that of choosing blue is 1/2. Since these events are occuring simultaneously and independently, the answer is 1/3 • 1/2 = 1/6.

Practice Questions 4.5

1 If the head of a coin is worth 1 point and the tail is worth 2 points, what is the probability that the sum of the results of 4 tosses is at least 5 points?

(A) 3/8
(B) 1/2
(C) 3/4
(D) 7/8
(E) 15/16

2 From a deck of 52 cards, 2 cards are randomly drawn. What is the probability that the first card is a 10?

(A) 1/26
(B) 1/17
(C) 1/13
(D) 12/51
(E) 1/4

3 A four-digit number is formed by randomly selecting from the digits 0 through 9. No digit may be selected more than once. If the first two digits selected are 8 and 6, what are the chances that the number is even?

(A) 1/64
(B) 3/64
(C) 1/24
(D) 1/21
(E) 3/8

4 If light switches numbered from 1 to 4 are either on or off in different combinations and switches 1 and 2 are always opposite (e.g., when switch 1 is on, switch 2 is off), how many different combinations are there if at least one light is on?

(A) 7
(B) 8
(C) 15
(D) 16
(E) 27

5

a, b, c, d

How many four letter arrangements, such as abcd, can be made using the letters above if the first letter must be a and no letters can be used more than once in the arrangement?

(A) 5
(B) 6
(C) 8
(D) 9
(E) 12

6

In a jar of 8 marbles, 3 are red. If you choose two marbles at random from the jar without replacement, what is the probability that both selections are red?

(A) 3/32
(B) 3/28
(C) 9/64
(D) 1/4
(E) 3/8

Questions 7-8 refer to the following scenario:

A singing contest has 10 contestants among whom
the judges must award 1st, 2nd, and 3rd prizes with no ties.

How many different ways can the judges award the three prizes?

(A) 180
(B) 240
(C) 360
(D) 494
(E) 720

8

If the judges can award 3 people each prize, how many different 3-person groups could receive one of the prizes?

(A) 180
(B) 240
(C) 360
(D) 494
(E) 720

9 The table below shows the distribution of a group of 100 employees of a certain company by gender and ethnicity:

	Hispanic	Caucasian	African American	Asian
Male	15	20	5	2
Female	18	18	8	14

If an worker is randomly selected from this group, what is the probability that the chosen employee is a female or Caucasian?

10 If the probabilities of Hank and Babe, two baseball players in the same MLB team, of hitting a home run in today's game is 0.3 and 0.2, respectively, what is the probability that at least one of them will hit a home run in today's game?

11 If 0, 1, 2, 3, and 4 are arranged to form a four-digit number, how many different even numbers are possible?

4.6 :: Statistics
Mean, Median, & Mode

try it yourself

Try this sample question within 30 seconds.

3, 2, 5, 3, 6, 2, 4, 4, 3, 3, 7, 7, 4, 3, 2

Based on the raw scores above, what is the median of the scores?

(A) 2 (B) 3 (C) 4 (D) 5 (E) 6

general rules

Numerical Description of Data: Statistics
Central Tendency: the arithmetic mean/average, median, and mode

☑ The **mean** (or arithmetic mean) is the average.
☑ The **median**, when all the values are listed in an ascending or descending order, is the value of the middle number when there is an odd number of values in the list, and the average of the two middle numbers if there is an even number of values on the list.
☑ The **mode** is the most frequently-occuring value.

☑ The **weighted mean from weighted sum** is used when some values are repeatedly occuring.

$$\frac{\text{(specific value)(100)}}{\text{total value}} = \frac{\text{(frequency)(value)(100)}}{\text{weighted sum}}$$

For example, in the statistical data: 1, 2, 3, 3, 4, 4, 5, 5, 5, 5 (in ascending order)

The **mean** is $\frac{1+2+3+3+4+4+5+5+5+5}{10} = 3.7$.

There is an even number of values (10).

The **median** is the average of the two middle numbers 4 and 4: $\frac{4+4}{2} = 4$.

The **mode** is 5 since it has the most frequent occurance of 4 times.

approach to sample questions

A1 List the scores in either ascending or descending order.
In ascending order, the scores are: 2, 2, 2, 3, 3, 3, 3, 4, 4, 4, 5, 6, 7, 7.
Because there is an odd number of values in the list, the median would be the number that appears in the middle of the list, which is in this case, the 8th score: 3. The answer is (B).

Practice Questions 4.6

Use the following data set for problems 1-4:

2, 2, 3, 2, x, 4, 3, 4, 4, 3, 3, 2, 4, 3, 4,
x is an integer score with range $1 < x < 5$.

Based on the above list of 15 numbers, answer the following.

1 What is the median score?

2 In order for the mode to be 4, what does x have to be?

3 If mean = median, what value must x be?

4 The table below shows how many students surveyed in a high school class have taken the SAT 0-3 times. Two students who were absent on the date of the survey were later added to the results (not shown below), making the average equal to the median. How many times did the two absent students take the SAT?

Number of times each student took the SAT	Number of students
0	1
1	6
2	8
3	8

Use the following data set for questions 5-6:

The SAT math scores of a class of 10 students are
610, 620, 620, 650, 650, 650, 740, 740, 750, and 770.

5 What is the median of the students' scores?

6 If an math drill increases each student's score by 30 points, what would be the difference between the new mean score and the new mode score of the 10 student scores?

Use the following data set for questions 7-8:

Three classrooms of students have the following mean and median numbers of college applications submitted this year.

	# of students	# of college applications mean	# of college applications median
Class A	20	8	6
Class B	30	6	8
Class C	10	8	6

7 What is the mean number of college applications submitted by the 60 students?

8 What is the median number of college applications submitted by each of the 60 students?

(A) 6
(B) 7
(C) 7.5
(D) 8
(E) Cannot be determined

9

4, m, 5, 9, 3, n

The arithmetic mean of the above sequence of numbers is 4. If m and n are non-negative integers and $m > n$, what is the median number in the list?

(A) 3
(B) 3.5
(C) 4
(D) 4.5
(E) Cannot be determined

4.7 :: Statistics
Data Interpretation/Analysis

try it yourself

Try this sample question within 30 seconds.

Q1 A certain school has 5 math teachers who teach algebra and geometry classes for 5 periods, shown in the table below. What fraction of the total teacher hours is alloted for geometry class during a 5-period day?

Class periods	# of math teachers allotted	
	Algebra	Geometry
1 and 2	2	3
3 and 4	1	4
5	3	2

general rules

Chart interpretation/analysis:
The following is a summary of the basic tools needed for interpreting data.

☑ **Weighted mean from weighted sum** = sum of (frequency)(value), when some values are repeatedly occuring.

☑ **Percent:** $\frac{\text{(specific value)(100)}}{\text{total value}}$ (simple %) $\frac{\text{(frequency)(value)(100)}}{\text{weighted sum}}$ (weighted %)

☑ **Table:** Indicates data through actual numbers. Calculate the sum, difference, or product across the rows or columns to find the total. If necessary, apply the weighted sum rule and create new rows/columns to solve the problem using these new, actual (weighted through multiplying by frequency) values.

☑ **Bar or Divided Bar graph:** Represents different categories in the x- or y-axis. Be mindful of the units and whether or not each axis starts at 0.

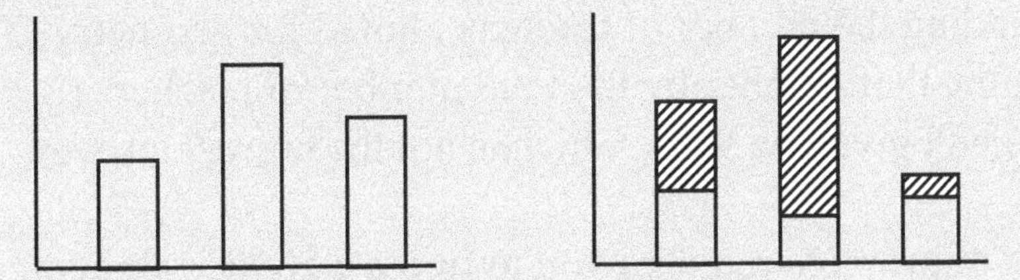

☑ **Histogram or Normal Distribution:** Represents the spread or distribution of data. Each unit is the same and each bar represents frequency.

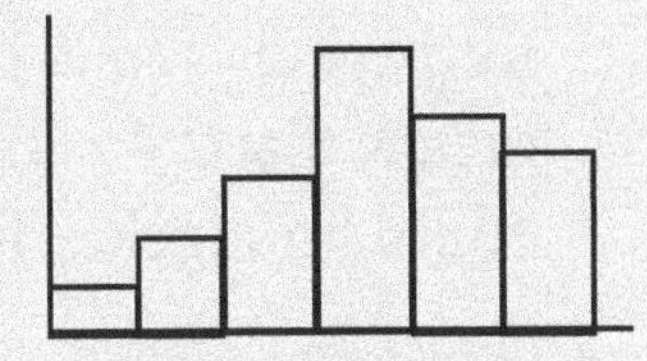

☑ **Pie chart:** 360° is equivalent to 100%. Given the total represented by the entire pie, find the actual value of each slice by multiplying the given percentage by the given total.

Total = 200
x = 50% = 100

Total = 400
x = 25% = 100

Though the areas of the two pies are different, the actual values of x are the same.

☑ **XY-Coordinate graph/Scatter plot:** Be able to identify both the x- and y-values for each coordinate - be wary of when they are switched! To separate the plots in question, use

a vertical line to find x
a horizontal line to find y
a diagonal line from the origin
(where the slope = the y/x ratio).

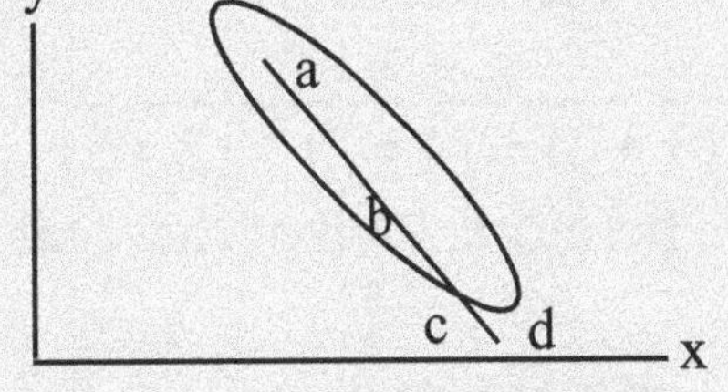

☑ **Time plot:** Represents changes of values (y-axis) over time (x-axis). The slope of the line indicates the amount/percent increase of a value between time periods. Compare multiple graphs at one point of time (draw a vertical line) to find the difference or percentage between the two values.

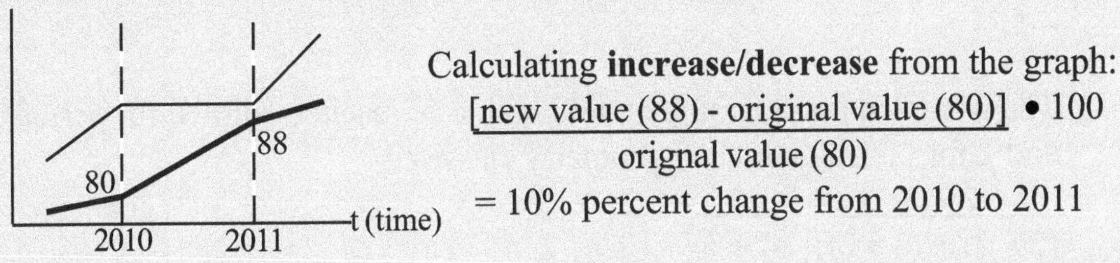

Calculating **increase/decrease** from the graph:

[new value (88) - original value (80)] • 100
orignal value (80)

= 10% percent change from 2010 to 2011

approach to sample questions

A1 Simply adding the number of teachers alloted for geometry $(3 + 4 + 2 = 9)$ and dividing that number by the total number of teachers $(2 + 1 + 3 + 3 + 4 + 2 = 15)$ will give you 9/15, which is not the correct answer.

This type of problem requires you to find a weighted fraction in teacher hours; that is, you must see that if there are 2 teachers allocated for Algebra for periods 1 and 2, there are actually $2 \bullet 2 = 4$ teacher hours total, which means that $2/4 = 1/2$ is the weighted fraction of the number of teacher hours for algebra for periods 1 and 2.

We can reconstruct the given table to reflect the data in weighted form. Multiplying the number of teachers by the number of periods will give us the weighted results, which are presented in parentheses:

# of periods in a day	# of math teachers allotted		Total # of teachers
	Algebra	Geometry	
2 (1 and 2)	2 (4)	3 (6)	5 (10)
2 (3 and 4)	1 (2)	4 (8)	5 (10)
1 (5)	3 (3)	2 (2)	5 (5)
Total: 5 periods	5 (9)	9 (16)	

For the total of five periods, there are

$$\frac{3 + 4 + 2}{(3 \bullet 2) + (4 \bullet 2) + (2 \bullet 2)} = \frac{9}{6 + 8 + 2} = \frac{9}{16}.$$

Therefore, 9/16 of the total teacher hours has been alloted for geometry.

For more practice:

The weighted fraction of the number of teacher hours allocated for algebra in periods 3 and 4 is

$$\frac{1}{2 \bullet 1} = 2.$$

The weighted fraction of the total number of teachers alloted for periods 1 and 2 for both algebra and geometry classes is

$$\frac{2 + 3}{(2 \bullet 2) + (3 \bullet 2)} = \frac{5}{4 + 6} = \frac{5}{10} = \frac{1}{2}.$$

Practice Questions 4.7

Questions 1 & 2 refer to the following table:

One thousand people were surveyed for their opinions regarding the issue of abortion. For, against, and undecided about the issue are represented in the table below.

1000 people surveyed:

	For	Against	Undecided
Men	50%	40%	10%
Women	60%	20%	20%
Total (of 100)	53%	34%	13%

1 How many people are against abortion?

2 How many women were surveyed?

3 Which year has the highest class-student ratio?

(A) 1994
(B) 1995
(C) 1996
(D) 1997
(E) 1998

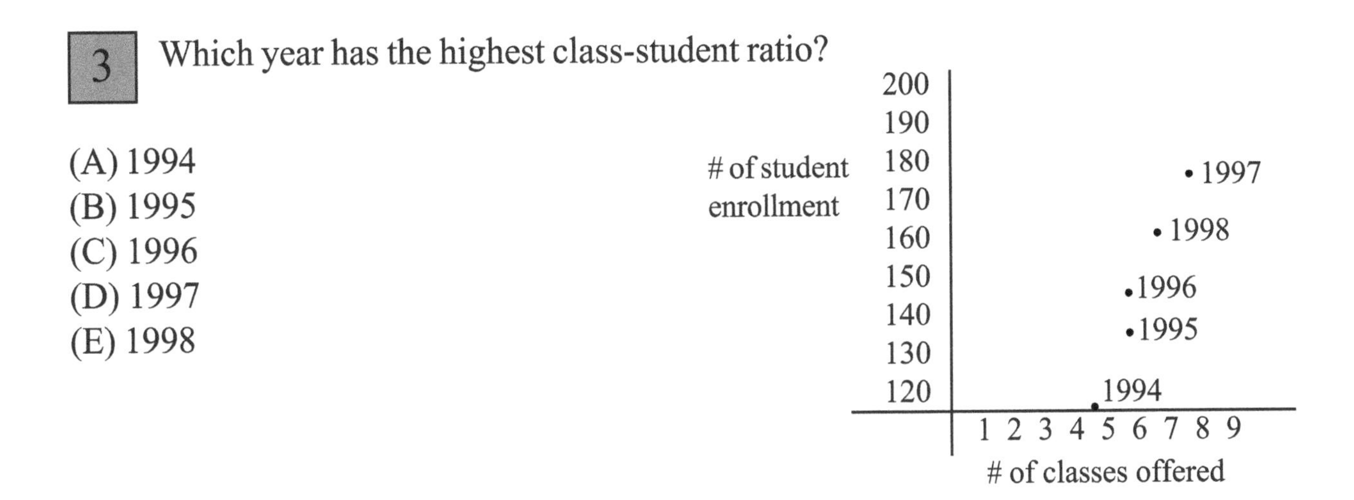

Questions 4-6 refer to the following charts:

% of students failing/withdrawn from classes

	Failing	Withdrawn
Algebra I	10%	5%
Geometry	20%	10%
Algebra II	20%	15%
Calculus	40%	40%

% students taking math classes

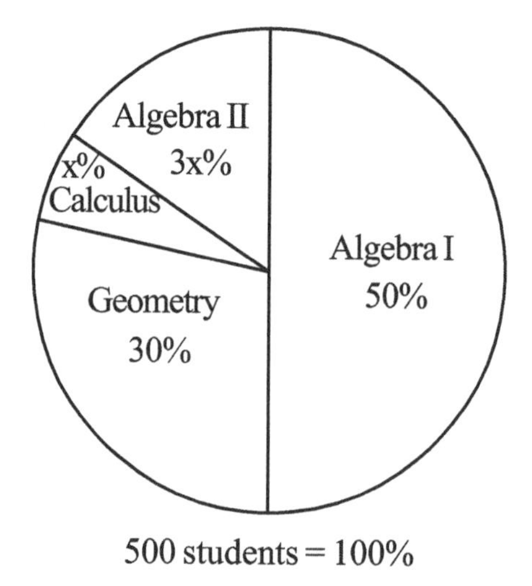

500 students = 100%

4 What is the difference between the number of students who withdrew from Geometry and the number of students who are failing in Calculus?

5 What percent of students are failing Algebra II?

6 What is the overall percentage of students failing across all the math classes?

7 Distribution of John's Family Expenses for the years 2009 and 2010

Category	2009	2010
Rent	30%	32%
Food	5%	4%
Utilities	5%	6%
Gasoline	2%	3%
Travel/Entertainment	8%	5%
Automobile	7%	7%
Medical	6%	8%
Credit	3%	4%
Education	20%	20%
Miscellaneous	14%	10%
Total	100%	100%
Total Income	**$50,000**	**$51,000**

According to the table above, which of the following statements is not true?

(A) Rent, medical, and educational expenses together comprise of more than 50% of the total expenses for both 2009 and 2010.
(B) Automobile expenses did not change from 2009 to 2010.
(C) John's family spent 2% more on gasoline in 2010 than it did in 2009.
(D) The amount increase in total income is less than the amount increase of money spent on rent between the two years.
(E) All of the statements are true.

8 The graph below displayes data from the US Department of Commerce. Which of the following statements is true?

(A) The GDP was highest in 2004.
(B) The GDP in 2003 was higer than in 2007.
(C) The GDP in 2008 was lower than in 2007.
(D) The GDP in 2010 was higher than in 2007.
(E) All of the statements are true.

9 The following pie graphs represent the distribution of Jason and Steph's total expenses for entertainment in 2010.

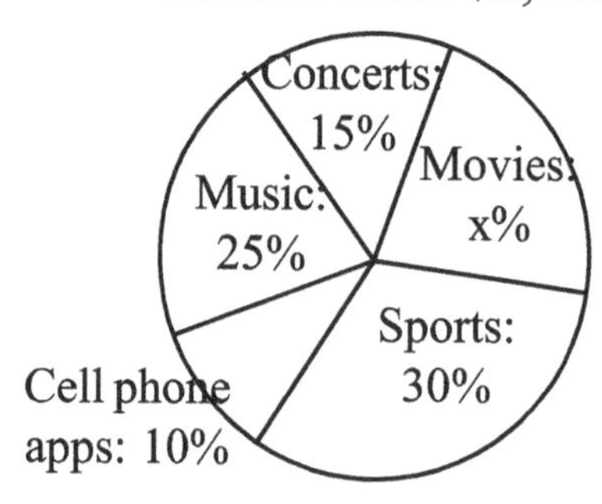

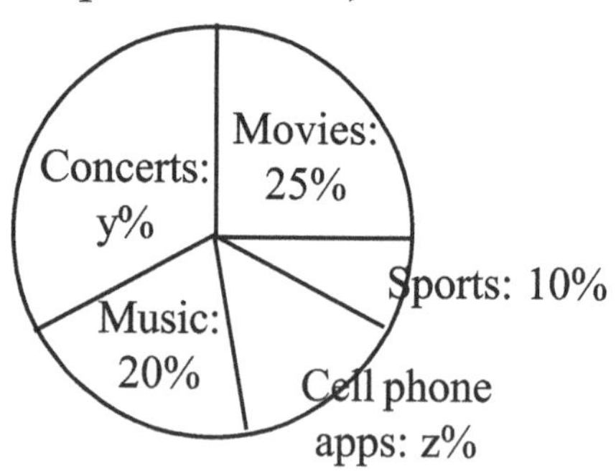

How much money did Jason and Steph spend on movies, concerts, and cell phone apps combined?

10 The divided bar graph to the right shows the breakdown of the yearly composition of David's income between his expenditures and savings for three different years. Based on this information, which of the following statements is not true?

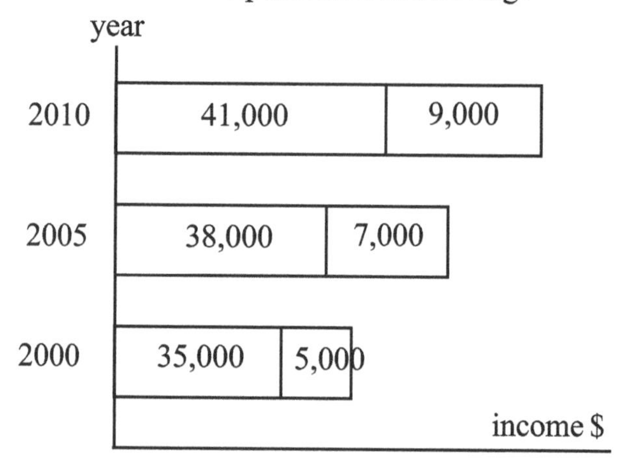

(A) There was a greater percent increase in savings than expenditures from the year 2000 to 2010.
(B) The savings/income ratio is the smallest in the year 2000.
(C) The percent increase in savings in the years 2000-2005 is greater than the percent increase in savings in the years 2005-2010.
(D) The increase in income for the years 2000-2005 is the same as in 2005-2010.
(E) All of the statements are true.

CHAPTER 4 TEST

1 If < x # > = the least integer greater than x, what is the result of the following mathematical operation?

< 3.2 # > - < -2.5 # >

(A) 1
(B) 2
(C) 3
(D) 4
(E) 6

2 If n is divided by 5, the remainder is 1. If 2n is divided by 5, what is the remainder?

(A) 0
(B) 1
(C) 2
(D) 3
(E) 4

3 The first two numbers on a certain list are 5 and 6. Starting with the third number on the list, each number is the sum of the two that immediately precede it. If the nth number on the list is the first to exceed 100, what is the value of n?

(A) 6
(B) 7
(C) 8
(D) 9
(E) 10

4 In a certain warehouse, there are fewer than 17 computers. If the computers are stacked 4 to a stack, there are three left. If they are stacked 3 to a stack there are two left. How many computers are in the warehouse?

(A) 11
(B) 13
(C) 15
(D) 17
(E) 19

5 A certain sequence of numbers has 2 as its first term. Every term after the first is 1 more than the cube of the term immediately before it. What is the third term of the sequence?

(A) 9
(B) 10
(C) 18
(D) 729
(E) 730

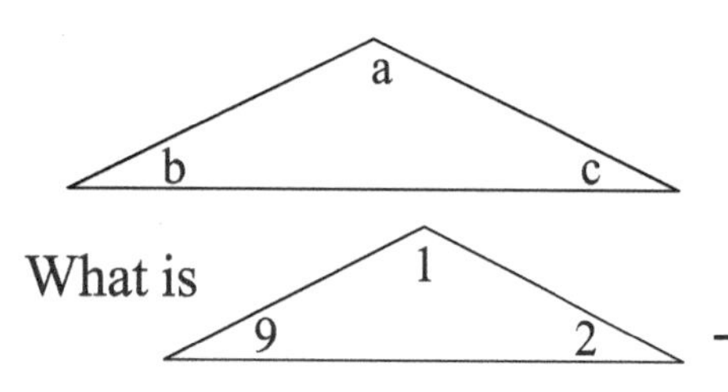

is the greatest prime factor for ab - ac.

What is

?

(A) 3
(B) 4
(C) 5
(D) 7
(E) 9

If x is the median, y is the mean, and z is the mode of a certain data-set consisting of 11 positive integers, and $x < y < z$, which of the following must be true?

(A) y is the arithmetic mean (average) of x and z.
(B) There are more integers that are less than x than that are greater than x.
(C) z is the greatest value among the data.
(D) If 2 integers of the lowest value are removed, the value of z will change.
(E) The sum of all the integers is 11y.

8 How many multiples of 7 are there between 35 and 693, excluding 35 and 693?

(A) 92
(B) 93
(C) 94
(D) 95
(E) 658

9 In a certain sequence, each term is generated by multiplying the previous term by 3, starting with 4. If the sum of the first k terms of the sequence is equal to 2 less than half the value of the (k + 1) term, and if the sum of the first k terms is 160, what is the value of k?

(A) 3
(B) 4
(C) 5
(D) 6
(E) 7

10

x, y, z, 1, 2, 3

How many six letter-number arrangements, such as xyz123, can be formed from the list above? The first three of the arrangement must be letters and the last three have to be numbers. No letter or number can be used more than once.

(A) 18
(B) 36
(C) 72
(D) 360
(E) 720

11 A class consists of students born in a certain year, where at least two people were born in each month of the year. What is the probability that two randomly-selected students were born in different months of the year?

(A) 1/6
(B) 1/3
(C) 1/2
(D) 5/6
(E) 11/12

12 X and Y are nonempty sets. If s is in set Y, then s is also in set X. Which of the following statements must be true?

I. If s is in set X, then s is in set Y.
II. If s is not in set Y, then s is not in set X.
III. If s is not in set X, then s is not in set Y.

(A) I only
(B) II only
(C) III only
(D) II and III only
(E) I, II, and III

13 The chart below shows the rate of students from a certain school passing and failing a driving test. Which of the following statements is true?

	Fail	Pass
10th grade	30%	70%
11th grade	60%	40%
Total	50%	50%

(A) There were more students in the 10th grade than in the 11th grade.
(B) There were more students in the 11th grade than in the 10th grade.
(C) There were the same number of students in both grades.
(D) Cannot be determined

The graph to the right is drawn to scale. If the number of births increased from b_1 to b_2 while the number of deaths increased from d_1 to d_2, what is not true about the change between 2000 and 2010?

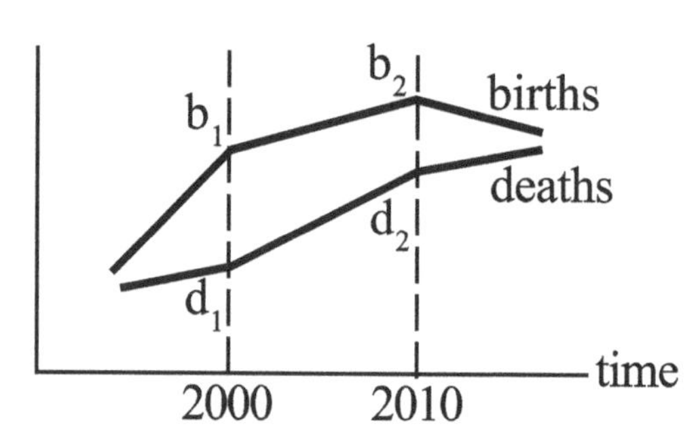

(A) The increase in the number of deaths is greater than the increase in the number of births.
(B) The number of births is greater than the number of deaths throughout the entire time period.
(C) The difference in the numbers of births and deaths in 2010 is approximately half of the difference in 2000.
(D) The rates of increase in births and deaths are reflected in the slopes of each line.
(E) All statements are true.

According to *Business Insider*, the US government owed about \$14.3 trillion in debt in 2011, \$4.5 trillion of which was owed to foreigners. The graph to the right depicts the percentage of each debt owed to various parties. What is not true about the nature of the US debt?

(A) About 68.5% of the total debt is owed to Americans.
(B) The US government borrowed more than 30% of its total debt from its Treasury and Social Security.
(C) More than 30% of the US total debt is owed to foreign countries.
(D) More than 10% of the US total debt is owed to China.
(E) All of the statements are true.

Word Problems

5.1 :: Conversion Formulas

try it yourself Try these three sample questions within 30 seconds.

For questions 1 - 3, let X be the unknown. Make equations for the following:

Q1 What is the present age of Terry, who is ten years older than John? Let x = John's age.

Q2 What is 5 less than 20% of the sum of a and b?

Q3 What is half the difference of x and y decreased by 60% of the sum of x and y?

general rules

Remember the **conversion formulas**, used to change verbal expressions into mathematical expressions.

☑ **Conversion formulas**:

what	x
is	=
of	•
sum/difference of x and y	$(x \pm y)$
more than/increased by	+
less than/decreased by	-

approach to sample questions

A1 Let X = John's age
Then Terry's current age = X + 10

A2 $X = 20\left(\frac{1}{100}\right)(a + b) - 5$

A3 $X = \frac{1}{2}(x - y) - 60\left(\frac{1}{100}\right)(x + y)$

Practice Questions 5.1

For questions 1 - 4, let X be the unknown. Make equations for the following questions:

1 Twice a number decreased by three times the number equals 5.

2 Half a number increased by the quantity of twice the number decreased by 30% of the number is equal to three less than the number.

3 Four times John's age five years ago is equal to 3 less than three times John's age two years from now. What is John's current age?

4 The sum of a number and its reciprocal is equal to the sum of two times the number and half the number, all divided by 5 more than the number.

5.2 :: One-Variable Approach

try it yourself Try these two sample questions within 30 seconds.

Q1 Last year, Mark had 3 more dates than three times the number of dates he had this year. If Mark had three dates last year, how many dates did he have this year?

Q2 John had three more speeding tickets than twice the number of tickets Mary received, and consequently lost his driving privilege. If the total number of tickets they both received is 18, how many tickets had John received before his driving privilege was revoked?

general rules

Remember the most fundamental and helpful setting in approaching any word problem:

☑ **One unknown:**
Let: X = the unknown
Equation:

☑ **Two unknowns:**
Let: X = the unknown (usually the latter in the question)
Then: the other known = ? in terms of x
Equation:

approach to sample questions

You can sometimes get the correct answer without approaching the question through the following method. However, if you want consistently correct answers, you should follow this procedure. Besides, you don't save that much time when you skip these simple steps.

One unknown:
Let: X = number of date requests Mark received this year
Equation: (# of date requests last year =) 3x + 3 = 3
3x = 0
x = 0

Two unknowns:
Let: X = the number of Mary's tickets (it is introduced later and can be used as a measuring stick)
Then: the number of John's tickets = twice the number of Mary's tickets + 3
= 2x + 3
Equation: Total = 2x + 3 + x = 18
2x + x + 3 = 18
3x = 18 - 3
3x = 15
x = 5 (number of Mary's tickets)
Then, the number of John's tickets = 2x + 3 = 13

Practice Questions 5.2

1 An apple has 4 times as many worms as an orange does. If the apple gave 5 worms to a pear, it would have 11 worms left. How many worms does the orange have?

2 If $x + 6$ is 7 more than y, then $x + 18$ is how much more than $y + 4$?

3 The weight of a stone plus 1/4 of its weight is equal to 15 pounds. What is the weight of the stone, in pounds?

4 In a lottery, Sarah won 5 dollars more than Jack, and Jack won 4 dollars more than Irene. If Irene won 3x dollars, how much did Sarah and Jack make together expressed in terms of x?

5 Wendy is older than Michael. If the average (arithmetic mean) of their ages equals the difference of their ages, what is the ratio of Wendy's age to Michael's age?

5.3 :: Consecutive Integers

try it yourself Try these two sample questions within 30 seconds.

Q1 Two brothers' ages are consecutive odd integers. If the sum of their ages is five more than the younger brother's age, what is the older brother's age?

Q2 If a, b, and c are consecutive odd integers and $a < b < c$, how do you express the value of $a + c$ in terms of b?

(A) $b + 4$
(B) $b + 2$
(C) $b + 6$
(D) $2b + 2$
(E) $2b$

general rules

☑ For the case of **regular consecutive integers**:

Let: $X =$ the first integer
Then: $X + 1 =$ the second integer
$X + 2 =$ the third integer
and so on
Equation:

☑ For the case of **even/odd consecutive integers**:

Let: $X =$ the first integer
Then: $X + 2 =$ the second integer
$X + 4 =$ the third integer
and so on
Equation:

approach to sample questions

A1 Remember, like consecutive even integers, consecutive odd integers also have common differences of 2, e.g. 1, 3, 5, and so on.
Therefore, Let: x = the younger brother's age
Then: $x + 2$ = the older brother's age
Equation: $x + x + 2 = x + 5$
$2x + 2 = x + 5$
$x = 3$ (the younger brother's age)
$x + 2 = 5$ (the older brother's age)

A2 Let: a = the first odd integer
Then: $b = a + 2$
$c = a + 4$
Equation: $a + c = 2a + 4 = 2(a + 2) = 2b$.
The answer is (E).

Practice Questions 5.3

1 If the sum of two consecutive even integers is 10, what is the value of the larger integer?

2 If a, b, and c are three consecutive integers, which of the following must be true?

I. At least one of these integers is divisible by 4.
II. One of these integers is 5.
III. At least two of these integers are divisible by 2.

(A) None
(B) I only
(C) II only
(D) III only
(E) I and III

3 The product of 5 consecutive integers is 0. What is the maximum possible value of their sum?

(A) 9
(B) 10
(C) 11
(D) 12
(E) 13

4 The sum of three consecutive positive integers is always divisible by

(A) 2
(B) 3
(C) 4
(D) 5
(E) 6

5 Four consecutive odd integers add up to 72. What is the average of the smaller three?

(A) 15
(B) 16
(C) 17
(D) 18
(E) 22

5.4 :: Digit

try it yourself Try these two sample questions within 30 seconds.

Q1 If S, a two-digit number, is added to the number formed when the digits of S are reversed, which of the following can be the result?

(A) 45
(B) 63
(C) 77
(D) 64
(E) 98

Q2 If a three-digit number formed with three nonzero digits is decreased by a value of a two-digit number that is formed by the first and last digits of the three-digit number, which of the following can *not* be the result?

(A) 120
(B) 150
(C) 170
(D) 180
(E) 240

general rules

☑ The general **digit problem** approach:

Let: t = tens' digit
u = units' digit

Then: the **number** = $10t + u$
the **reversed number** = $10u + t$

approach to sample questions

A1 Let: t = tens' digit
u = units' digit
Then: $S = 10t + u$
Reversed $S = 10u + t$

$$S + \text{reversed } S = 10t + u + 10u + t$$
$$= 11t + 11u$$
$$= 11(t + u)$$

Therefore, the answer has to be a multiple of 11.
The answer is 77 (C).

Let: h = hundreds' digit
t = tens' digit
u = units' digit
Then: the three-digit number = $100h + 10t + u$
the two-digit number is $10h + u$

Subtracting the latter from the former, the result is $90h + 10t = 10(9h + t)$. Since $9h + t$ has to be formed with nonzero values for h and t, the only impossible values for h and t is for the case of 180, where $h = 2$ and $t = 0$. All of the other cases give nonzero values for t. The answer is (D).

Practice Questions 5.4

1 A two-digit number has tens' digit A and units' digit B. What is the product of 3 and this number, in terms of A and B?

2 A two-digit number has tens' digit t and units' digit v. A three-digit number is formed by placing the digit m to the right of the two-digit number. If the three-digit number is subtracted from ten times the two-digit number, the resulting number can be represented by

(A) $10t - m$
(B) $10t + v + m$
(C) $10t + m$
(D) $-m$
(E) $100t - m$

3 If the digits of a two-digit number S are reversed and the number obtained is subtracted from S, then the result could be any of the following EXCEPT

(A) 18
(B) 45
(C) 63
(D) 64
(E) 72

4 In the correctly-solved algebraic addition below, which of the following can be an answer?

$$\begin{array}{rr} & ABA \\ + & BAB \\ \hline \end{array}$$

(A) 325
(B) 437
(C) 749
(D) 866
(E) 999

5.5 :: Age

try it yourself

Try these two sample questions within 30 seconds.

Q1 The sum of Nicole's and her boyfriend's present ages is 31. Fourteen years ago, Nicole's boyfriend's age was twice her age. What is Nicole's present age?

Q2 In h years, John will be 3k + 1 times his current age. What is John's current age in terms of h and k?

(A) $2h/(3k + 1)$
(B) $h/(3k + 1)$
(C) $(2h - 1)/3k$
(D) $h/3k$
(E) $h(3k + 1)$

general rules

Let: x = one person's unknown present age
Then: (some x) = the other person's present age

	k years ago	Now	h years later
One person	x- k	x	x + h
Other person	(some x) - k	some x	(some x) + h

approach to sample questions

A1 Whenever the total, N, is given as the sum of two unknowns, it can be separated into two variables: x and N - x.

Let: x = Nicole's present age
Then: $31 - x$ = Boyfriend's age
Equation: $31 - x - 14 = 2(x - 14)$
$17 - x = 2x - 28$
$45 = 3x$; $x = 15$ (Nicole's present age)
$31 - x = 16$ (Nicole's boyfriend's age)

	14 years ago	Now
Nicole	$x - 14$	x
Boyfriend	$(31 - x) - 14$	$31 - x$

A2 $x + h = (3k + 1)x$
$x + h = 3kx + x$
$x = h/3k$
The answer is (D).

	Now	h years later
John	x	$x + h$

Practice Questions 5.5

1 A woman was 27 years old when her child was born. When the woman is four times as old as her child, how old will the child be?

2 Today Jack is twice as old as Mary was 9 years ago. How old is Mary today in terms of Jack's current age, Y?

(A) Y + 9
(B) Y + 18
(C) (Y + 9)/2
(D) (Y + 18)/2
(E) (Y/2) - 9

3 Benjamin is now 8 years older than Jeremiah. In five years, Benjamin will be twice as old as Jeremiah will be. How old is Benjamin now?

4 Sandy is T years old, which is 4 times Sharon's age. In terms of T, after how many years will Sandy be twice as old as Sharon?

(A) T/4
(B) T/3
(C) T/2
(D) T
(E) 2T

5 Exactly ten years ago Marge was 25 years old. Exactly 20 years ago, Homer was 20 years old. How much more will Homer's age 5 years from now be than Marge's current age?

5.6 :: Proportion

try it yourself

Try these three sample questions within 30 seconds.

Q1 James spent 180 minutes reading 15 pages of a book assigned in his English class. At this rate, how many hours does he need to finish this 600-page book?

Q2 If James spends 5 hours reading each day, it takes 12 days to finish an entire book. If he spends 2 hours a day, how many days does it take to finish the book?

Q3 If Jane makes h number of products in t hours, how many products does she make in x hours?

general rules

☑ **Direct proportion**

$$\frac{a}{b} = \frac{A}{B}$$

a and b vary in the **same** direction

☑ **Inverse proportion**

$$a \bullet b = A \bullet B$$

a and b vary in the **opposite** direction

There must be an agreement of units among the same variables.
For example, if a is in hours, then A also must be in hours.

☑ **Work type:** # of workers and time required to finish a certain task
☑ **Consumption type:** # of consumers and time required to consume
☑ **Wheel type:** size of wheel and # of revolutions
☑ **Fulcrum type:** weight and distance from the center of the fulcrum

approach to sample questions

A1 First, convert 180 minutes into 3 hours.
Then use the direct proportion formula as follows:

Let: $x = \#$ hours

Equation: $3 \div 15 = x \div 600$

$x = 3 \bullet 600 \div 15$

$x = 120$ hours

A2 Notice the inverse relationship and use the following formula:

Let: $x = \#$ of days required to finish the book

Equation: $5 \bullet 12 = 2 \bullet x$

$x = 30$ days

A3 Using the direct proportion formula,
Let $j = \#$ of products j makes in x hours.

$h/t = j/x$

$j = hx/t$

Practice Questions 5.6

1 If an airplane travels at a constant rate of 50 miles per second, how many minutes will it take to fly 600 miles?

2 A giant casts a shadow 72 meters long. A tree standing beside the giant is 3 meters tall and casts a shadow 12 meters long. How tall is the giant?

3 A weight of x pounds is equal to approximately n grams. A weight of p pounds is equal to approximately how many grams?

(A) xp/n
(B) xpn
(C) x/np
(D) xn/p
(E) np/x

4 If a quart of punch contains 1/2 pints of soda, then 1/4 pints of the same punch contains how much soda? (1 quart = 2 pints)

5 If an 8 x 4 carpet costs $500, how much will a 24 x 2 carpet cost?

5.7 :: Ratio

try it yourself Try these two sample questions within 30 seconds.

Q1 Acccording to K's school census, the ratio of various ethnic groups of the student body is as follows:
Caucasian : Hispanic : African-American : Asian = 6 : 4 : 3 : 2.
What part of the student body is Asian? If the total student population is 300, how many students are Asian?

Q2 The slam dunk contest result was 5 to 6 between Michael and Magic and 9 to 11 between Michael and Shaq. Who was the winner among the three?

general rules

☑ If the ratio among a, b, c, and d is k : l : m : n,
First, **find the total** of all the ratios: k + l + m + n.
Then, a is $\frac{k}{k+l+m+n}$ part of the total k + l + m + n.
If the total is given as W, the actual value of a is $\frac{(W) \bullet k}{k+l+m+n}$

☑ If a : b = 1 : 2 and b : c = 3 : 5, then a : b : c can be figured out as follows:

```
  a : b : c
  1 : 2
  ╲ | ╲
x     3 : 5
  ---------
  3 : 6 : 10
```

approach to sample questions

A1 Asian students are 2 parts of the total (6 + 4 + 3 + 2), that is, 2/15.
Therefore, $\frac{2}{15} \bullet 300 = 40$ Asian students.

A2

Michael	:	Magic	:	Shaq
5	:	6		
x 9			:	11
45	:	54	:	55

Therefore, Shaq is the winner.

Practice Questions 5.7

1 If 3/5 of the marbles in a box is red, what is the ratio of the number of red marbles to the number of those that are not red?

2 A recipe calls for x cups of oil, y cups of sugar, and z cups of salt. What part of the complete recipe is sugar?

(A) $x / (y + z)$
(B) $y / (x + z)$
(C) $1 / xz$
(D) $x / (x + y + z)$
(E) $y / (x + y + z)$

3 A baseball team has a win-to-loss ratio of 5 to 4. If the team has played a total of 27 games and no game has ended in a tie, how many more games has the team won than it has lost?

4 If the ratio of m to n is 5 to 7 and the ratio of m to p is 4 to 5, what is the ratio of n to p?

5 At a certain park, people are divided into four general age groups: infants, children, adolescents, and adults. The ratio among them is 3 : 15 : 12 : 20. If the total number of people at the park is 150, how many children and adolescents are there?

5.8 :: Average

try it yourself Try these two sample questions within 30 seconds.

Q1 Jane's average score of 4 exams in mathematics is 56, and she failed to take a recent test because she was ditching school with her friends.There is only one exam left before her final grade is to be determined. What is the lowest score Jane can receive in the exam in order to pass the class with a D (at least an average of 60)?

Q2 The average of s, t, u, v, and w is 40, and the average of v and w is 70. What is the average of s, t, and u?

general rules

☑ General strategy: **Get the total!**
Set Total = Total

☑ Get the total from the information about the partial average.
Total = (# items) • (average of these items)

For example,
If the average of 5 tests is 80, the total is 5 • 80 = 400.

approach to sample questions

A1 In average problems, always find the total first!
The total of four exams, the score of zero, and the potential score from the sixth exam is equal to the total score for all six exams.

Let: x = potential score from the sixth exam

Then: $4 \bullet 56 + 0 + x = 6 \bullet 60$

$224 + x = 360$

$x = 136$

Therefore, unless there are lots of extra credit projects, there is no hope for Jane.

The total, according to the formula, is $s + t + u + v + w = 40 \bullet 5 = 200$.
$v + w = 70 \bullet 2 = 140$.
Therefore, $s + t + u = 200 - 140 = 60$.
Dividing both sides by 3 yields $(s + t + u) / 3 = 20$.

Practice Questions 5.8

1 The average of three numbers is ten. If the sum of two of the numbers is 16, what is the third number?

2 If the average of a, b, and c is 20, what is the value of $a + b + c$?

3 The average of 4 numbers is 17. If one of the numbers is 3, what is the value of the sum of the other three numbers?

4 The average of three numbers is 10. If the average of y and these 3 numbers is 12, what is the value of y?

5 The average of Peter's Math and English grades is 90, while the average of his Science, Spanish, and P.E. grades is 80. What is Peter's average grade across all five subjects?

5.9 :: Percent

try it yourself Try these two sample questions within 30 seconds.

Q1 70% of the students who attended prom were only 20% of the 350 students in the senior class. How many attended prom?

Q2 Ryan, a high school sophomore, had a certain number of friends who would ask him for rides in his old hatchback car. After buying a brand new sports car, the number of Ryan's riding buddies increasd by 60%. However, after an accident, that number dropped by 75%, with 6 remaining friends still wanting a ride in his damaged car. How many friends asked Ryan for rides in his old hatchback?

general rules

☑ **Direct conversion problem**: translate using the **conversion formulas** (see Section 5.1):

20 is what percent of 70?

$$20 = \frac{X}{100} \bullet 70$$

☑ **Increase type problem**: to get the result after incorporating increase, profit, or interest of r %:

Multiply the original amount by $\frac{100 + r}{100}$

☑ **Decrease type problem**: to get the result after incorporating decrease, loss, or discount of r %:

Multiply the original amount by $\frac{100 - r}{100}$

approach to sample questions

A1 Let: x = # of attending students

Equation: $\frac{70x}{100} = \frac{20}{100} \cdot 350$

$x = \frac{20}{100} \cdot 350 \cdot \frac{100}{70} = 100$ students

A2 Use the increase/decrease formula introduced in this section. You'll never fail in any kind of percent problem.

Let: x = original number of friends

Equation: $6 = x \cdot \frac{(100 + 60)}{100} \cdot \frac{(100 - 75)}{100}$

$6 \cdot \frac{100}{160} \cdot \frac{100}{25} = x$

$x = 15$

Practice Questions 5.9

a. Percent: Direct Conversion

1 25 percent of 300 is equal to 75 percent of

2 If X is 150 percent of Y, then Y is what percent of X?

3 A soccer team won 10 games and lost 5 games. If the team wins the next k games, it will have won 75 percent of all the games played. What is the value of k?

4 If on a certain day there were only p of 32 students present for a computer class, the number of students absent was what percentage of the number of students present, expressed in terms of p?

(A) $\frac{32 - 100p}{p}$
(B) $\frac{p}{100 - 32p}$
(C) $\frac{100(p - 32)}{32}$
(D) $\frac{100(32 - p)}{p}$
(E) $\frac{p(32 - p)}{100}$

5 As a fee for his service, a stockbroker kept 20 percent of the total profit he earned from investment. If the customer received a divident of $1, how much, in dollars, did the stock broker earn from his service?

(A) $0.80
(B) $1.20
(C) $1.25
(D) $1.50
(E) $1.80

b. Percent: Increase/Decrease

1 The price of a suit that regularly sells for \$k is reduced by 40 percent, and then increased by 40% of the reduced price. What is the total percent decrease in the suit's price?

2 If the length and width of rectangle A are respectively 10 percent less and 20 percent less than the length and width of rectangle B, the area of A is equal to what percentage of the area of B?

3 If x is 10 percent greater than n and y is 30 percent greater than m, then xy is what percent greater than nm?

4 A shoe vendor reduced the marked price of a shoe by 50% and was still making a profit of 10% from the original cost. If x is the original cost, what is the marked price in terms of x?

(A) 1.1x
(B) 2.2x
(C) 3.3x
(D) 4.4x
(E) 5.5x

5.10 :: Motion

try it yourself

Try this sample question within 60 seconds.

Q1 On an 3 mile running trail in Yosemite National Park, Terry ran at 8 mph for the first 2 miles, and at 4 mph for the last mile. On her way back, she walked at 3 mph for the entire 3 miles. How long did this round trip take? What was her overall average rate of travel?

general rules

☑ **Fundamental formula**: Rate • Time = Distance

R • T = D **R = D/T** **T = D/R**

☑ **Normal trip in the same/opposite direction**:

Let: x = unknown

Then: fill in the blanks in terms of x

i) If x = R, T = D/x

ii) If x = D, R = x/T

	R	T	D
case 1	x	D/x	D
case 2	x/T	T	x

Equation:

☑ **Round trip with/against a current**:

Let R = rate of the boat

r = rate of the current

	R	T	D
with current	R + r	t_1	D_1
against current	R - r	t_2	D_2

Equation: $(R + r)t_1 = D_1$

$(R - r)t_2 = D_2$

☑ Be careful when calculating the **average speed**: this is not the average of two different speed rates

$$\text{Average rate} = \frac{\text{Total distance}}{\text{Total time}}$$

approach to sample questions

A1 For any motion problem, draw a chart and fill in the blanks according to the given information.

In this problem, we are given Terry's distance and speed for each way of her trip. Since R • T = D, T = D/R. We can thus fill in the T column of the table.

	R	T = D/R	D
1-way a	8	2/8 = 0.25	2
1-way b	4	1/4 = 0.25	1
Return	3	3/3 = 1	3

The trip took Terry 0.25 + 0.25 + 1 = 1.5 hours.

$$\text{Average speed of travel} = \frac{\text{total distance}}{\text{total time}} = \frac{2+1+3}{0.25+0.25+1} = \frac{6}{1.5} = 4 \text{ mph}$$

Practice Questions 5.10

1 A kangaroo left the zoo at 12:00 noon and reached the Sydney Opera House at 2:30 p.m. the same day, having hopped a distance of 400 kilometers. What was the kangaroo's average speed in kilometers per hour?

2 A boat traveling upstream at a rate of 30 miles per hour is moving against a 10 mph current. If the boat travels up the 40-mile stream and returns the same distance, how long will the complete round trip take?

3 A man leaves a place at 11 a.m. traveling at a rate of 40 miles per hour. A faster man leaves the same place at 1 p.m. that afternoon and travels in the same direction on the same road at a rate of 60 miles per hour. At what time will the faster man overtake the slower man?

4 The distance from S to U via T is 40 miles. If one travels from S to T at a rate of 10 miles per hour, and from T to U at a rate of 30 miles per hour, the trip takes exactly 2 hours. What is the distance from S to T?

5 A plane flew 300 miles at 600 mph and returned the same distance at 300 mph. What was the average speed of the plane for the round trip?

5.11 :: Mixture

try it yourself

Try these three sample questions within 90 seconds.

Q1 At a fundraising party of 15, Jim collected \$51. If the entrance fee was \$2 for women and \$5 for men, how many men and women attended the party?

Q2 Jim mixed 2 gallons of soda (\$1/gallon) and 3 gallons of fruit punch (\$2/gallon) to make five gallons of Shirley Temples. If one gallon holds ten cups of the drink, how much must he charge per cup to break even?

Q3 If at a party a person drinks 2 cups of beer containing 6% alcohol and 1 cup of scotch containing 45% alcohol, is he allowed to drive right after the party, assuming that the legal limit is 0.08% blood alcohol level regardless of the person's weight?

general rules

☑ **General formula for a mixture problem:**
$AX + BY = (A + B)Z$

☑ **Coin / Stamp / Ticket mixture:**
Total # of coins & total $ value are usually given
Let: x = # of one type (dimes, worth 10)
Then: T (total) - x = # of the other type (quarters, worth 25)
Equation: $x \bullet 10 + (T - x) \bullet 25$ = total $ value

☑ **Coffee / Candy mixture and its average price:**
If the total amount is given, Let: x = amount of type 1
Then: T - x = amount of type 2
Equation: $x \bullet P_1 + (T - x) \bullet P_2 = T \bullet P_{mix}$
If the amount of each kind (x & y) and respective unit prices (P) are given,
Equation: $x \bullet P_1 + y \bullet P_2 = (x + y) \bullet P_{mix}$

60% vinegar	100% total
40% oil	

☑ **Solution mixture:**
If the total amount is given, Let: x = amount of type 1
Then: T - x = amount of type 2
where R is % of the substance of interest
Equation: $x \bullet R_1 + (T - x) \bullet R_2 = T \bullet R_{mix}$
If the total amount is not given, but the amount (X & y) and corresponding R(%) are given,
Equation: $x \bullet R_1 + y \bullet R_2 = (x + y) \bullet R_{mix}$

approach to sample questions

Since the problem is a ticket mixture and the total number is given,
Let: x = # of women
Then: 15x = # of men
Equation: $x \bullet 2 + (15 - x) \bullet 5 = 51$
$2x + 75 - 5x = 51$
$24 = 3x$
$x = 8$ (women)
$15 - 8 = 7$ (men)

A2 The question asks for the average price of a mixed drink.
Let: P = price of the mix
Equation: $x \bullet P_1 + y \bullet P_2 = (x + y) \bullet P_{mix}$
$2(1) + 3(2) = (2 + 3)P$
$8 = 5P$; $P = \$1.6$ per gallon and \$.16 per cup.

A3 The question deals with a solution mixture.
Let: R = % alcohol in the person's blood
Equation: $x \bullet R_1 + y \bullet R_2 = (x + y) \bullet R_{mix}$
$2(6) + 1(45) = (2 + 1)R$
$57 = 3R$; $R = 19$ (% alcohol)
It is against the law if this person drives.

Practice Questions 5.11

1 Michael has a combination of nickels and quarters which amounts to \$1.50. If the total number of coins is 10, how many nickels and quarters does he have?

2 A candy dealer makes up a 7-pound mixture of candy by combining 3 pounds of candy at 30 cents per pound, 2 pounds of candy at \$1.50 a pound, and 2 pounds of candy at 50 cents per pound. What is the price of this mixture of candy per pound?

3 How many gallons of water must be added to 20 gallons of a 10% solution of salt and water to reduce it to a 4% solution?

4 In a barrel, there are 50 pints of a solution that is 20% sugar. How many pints of pure sugar must be added to produce a solution that is 40% sugar?

5 In a 10-kilogram solution of water and oil, the ratio by mass of water to oil is 3 : 2. If 8 kilograms of a solution consisting of 2 parts of water to 1 part of oil are added to the 10-kilogram solution, what fraction by mass of the new solution is oil?

5.12:: Work

try it yourself

Try these two sample questions within 30 seconds.

Q1 It takes 6 people 2 hours to solve 3 problems. If 4 people work together for 1 hour, how many problems can be solved?

Q2 When Michael works alone it takes 3 hours to finish his homework project, while for Jenny it takes 2 hours to finish the same work. How many more hours would it take for Michael to finish if he works alone for 1 hour and then calls Jenny to help him?

TIP: When a person takes k days to finish a task, his/her rate of work is 1/k. In the work problem setting, a completed job is 1.

general rules

☑ **Multi-production approach**

$$\frac{m_1 \bullet t_1}{p_1} = \frac{m_2 \bullet t_2}{p_2}$$

where "m" workforce takes "t" days to produce "p" products
(m:t = inverse; m:p = direct proportion)

If different rates of workforce m's are used, convert them into the same unit of workforce to compare their rate of work.

☑ **Combined work approach**

$$(1/t_1 + 1/t_2) \bullet T = 1$$

Combined rate • time = work

Example: Person 1 takes t_1 days and Person 2 takes t_2 days to finish a certain task.

approach to sample questions

A1 Using the multiproduction approach,

$$\frac{a \bullet b}{c} = \frac{A \bullet B}{C}$$

Let x: = # of problems solved

Equation: $\frac{6 \bullet 2}{3} = \frac{4 \bullet 1}{x}$

$x = 1$ (one problem)

A2 Using the combined work approach,

Rate • Time = 1 (complete work)

Let: x = # extra hours taken

Equation: $(1/3)(1) + (1/3 + 1/2)x = 1$

$2 + 2x + 3x = 6$

$5x = 4; x = 4/5$ (hours)

Remember, when you work with an equation with fractions, multiply every term by the L.C.D. to make it easier to handle. In this case, multiply every term by 6.

Practice Questions 5.12

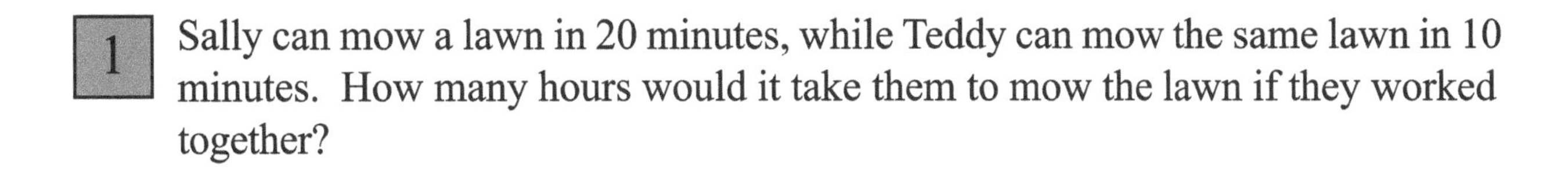

1 Sally can mow a lawn in 20 minutes, while Teddy can mow the same lawn in 10 minutes. How many hours would it take them to mow the lawn if they worked together?

2 6 tractors working together can plow a field in 12 hours. How long would it take 8 tractors to plow a field of the same size, if all tractors work at the same rate?

3 A worker can produce half of a product in one day. At this rate, how many products can three workers finish in 4 days?

4 A machine of type R can produce 40 bolts in x minutes. How many minutes would it take for two machines of type R to produce 120 bolts?

5 John can finish a job in 5 hours, while Liz can finish the same job in 3 hours. If they work together for an hour and John finishes the rest of the job alone, how many hours would it take for John to finish the rest of the job?

CHAPTER 5 TEST

1 If c is not equal to zero, and the value of (a + 3c) exceeds 2c by 150 percent of c, then a/c =

(A) -7/2
(B) -2/7
(C) 1/2
(D) 3/2
(E) 2

2 If v, w, x, and y are consecutive odd integers, and $v < w < x < y$, then x + y is how much greater than v + w?

(A) 2
(B) 4
(C) 6
(D) 8
(E) 10

3 If the digits of a two-digit number T are reversed and the number obtained is subtracted from T, then the result can be any of the following EXCEPT

(A) -72
(B) -45
(C) 0
(D) 27
(E) 64

4 Mary's age is half of Paul's age. In 8 years, Paul will be n years old. In terms of n, how old is Mary now?

(A) (n/2) - 4
(B) n/2
(C) (n/2) + 4
(D) (n/2) + 8
(E) 2n + 8

5 A student's average on the first four tests in an algebra course was 60. After taking the fifth test, the student's average was 65. What was the student's grade on the fifth test?

(A) 81
(B) 85
(C) 88
(D) 90
(E) Cannot be determined

6 An automobile can travel (m + 1) miles on g gallons of gasoline. At this rate, how many gallons of gasoline would it need to travel m miles?

(A) g - 1
(B) (g - 1) / m
(C) g / m
(D) 8 / (m + 1)
(E) gm / (m + 1)

7 During a certain month, for every 2 sales of movie V, there were 5 sales of movie S. If the total number of sales of movies V and S for that month was 280, how many of those were sales of movie S?

(A) 80
(B) 100
(C) 160
(D) 200
(E) 240

8 If 30 percent of 4x is multiplied by 20 percent of x, the result is what percent of $(2x)^2$?

(A) 2%
(B) 4%
(C) 6%
(D) 12%
(E) 25%

9 Which of the following changes in a positive number n is equivalent to decreasing n by 10 percent and then increasing the result by 40 percent?

(A) Decreasing n by 10%
(B) Decreasing n by 2%
(C) Increasing n by 6%
(D) Increasing n by 26%
(E) Increasing n by 30%

11 One cup of salt and one cup of water are added to two cups of an original salt-and-water solution. If the yield is 4 cups of solution that is 1/2 salt, what fraction of the original solution was salt?

(A) 1/6
(B) 1/4
(C) 1/3
(D) 1/2
(E) 2/3

12

x, y, 22, 46, 94

In the sequence above, each term after the first term, x, is formed by doubling the previous term and then adding 2. What is the value of x + y?

(A) 10
(B) 12
(C) 14
(D) 16
(E) 18

13 If the sum of 6 and a certain number x is equal to 10 minus the same number x, then x equals

(A) 0
(B) 2
(C) 4
(D) 6
(E) 8

14 The sum of 3 consecutive odd integers is x. What is the smallest of these 3 integers, in terms of x?

(A) x/3
(B) (x + 1)/3
(C) (3x - 4)/3
(D) x/3 - 6
(E) (x - 6)/3

15 A two-digit number has a tens' digit x and a units' digit y. What is the product of this number and the number 8, in terms of x and y?

(A) 8x + y
(B) 8x + 8y
(C) 8x + 80y
(D) 80x + 80y
(E) 80x + 8y

16 The present ages, in years, of 3 sisters are three consecutive even integers. Two years ago, the sum of their ages was 30. What is the present age of the youngest sister, in years?

(A) 6
(B) 8
(C) 10
(D) 12
(E) 14

17 If the average of 10, 20, m, and n is 50, the average of m and n is

(A) 65
(B) 70
(C) 80
(D) 85
(E) 95

18 If the manager of a store makes telephone calls at the rate of x calls every t work-days, what is the number of telephone calls the manager makes in y work-days?

(A) xy/t
(B) x/yt
(C) yt/x
(D) y/xt
(E) t/xy

19 Of the 60 students in a room, the ratio of girls to boys is m to n. How many girls are in the room in terms of m and n?

(A) 60m
(B) (m + n) / 60m
(C) 60m / n
(D) 60 / mn
(E) 60m / (m + n)

20 If 40 percent of x is equal to y, then, in terms of y, 40 percent of 4x is equal to

(A) y/16
(B) y/4
(C) y
(D) 4y
(E) 12y

21 If v is 10% greater than x and w is 20 percent greater than y, then vw is what percent greater than xy?

(A) 20%
(B) 30%
(C) 32%
(D) 50%
(E) 600%

A car traveled 600 miles one way at 60 mph and returned the same distance at 40 mph. What was the average speed of the car?

(A) 45 mph
(B) 48 mph
(C) 50 mph
(D) 52 mph
(E) 55 mph

23 If 15 kilograms of pure water is added to 25 kilograms of pure alcohol, what percent, by weight, of the resulting solution is alcohol?

(A) 10%
(B) 25%
(C) 40%
(D) 62.5%
(E) 66.67%

24 If 2 liters of water evaporate from 12 liters of 30% salt solution, what is the resulting % of the solution?

Special Types of Word Problems

6.1 :: Set (Venn Diagram)

try it yourself Try these two sample questions within 60 seconds.

Q1 Among Ali's class of 40 students, 13 people speak Spanish, 30 people speak English, and 6 people speak both English and Spanish.
1) How many people speak neither language?
2) How many people speak only one language?

Q2 Of 300 students in Robert's school, 2/3 have PC's and 60 have Macs. Of the students who have PC's, 30 have Macs as well. How many students have neither PC's nor Macs?

general rules

☑ **Venn diagram** approach:
If a person or thing belongs to group A or group B,

Total = m(A or B) + m(neither A nor B)
m(A or B) = m(A) + m(B) - m(A and B)
m(A) + m(B) = m(A or B) + m(A and B)

where m represents members.

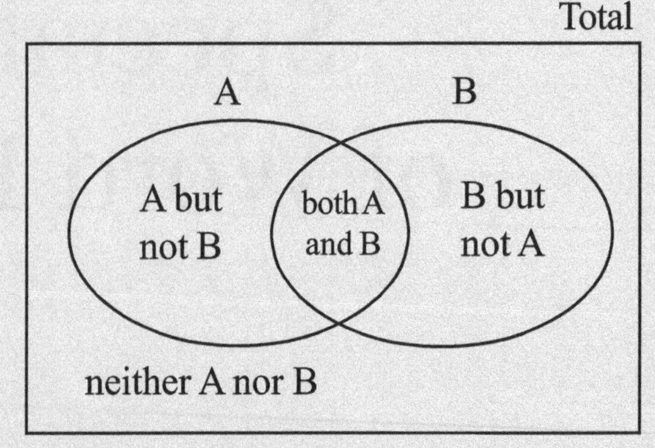

approach to sample questions

A1 1) m(A or B) = m(A) + m(B) - m(A and B)
= 13 + 30 - 6 = 37
m (neither A nor B) = total - m(A or B)
= 40 - 37 = 3
2) Only one language = 7 + 24 = 31

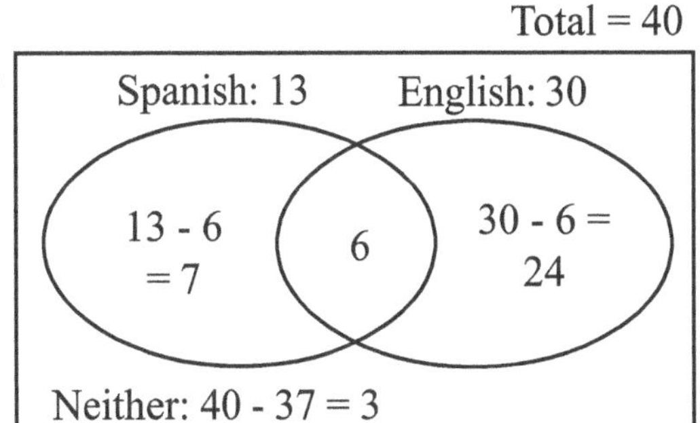

A2 Total = 300
m(A) = PC owners = 2/3 of 300 = 200
m(B) = Mac owners = 60
m(A and B) = owner of both = 30
m(A or B) = computer owners
= m(A) + m(B) - m(A and B)
= 200 + 60 - 30 = 230
Students without computers = 300 - 230 = 70

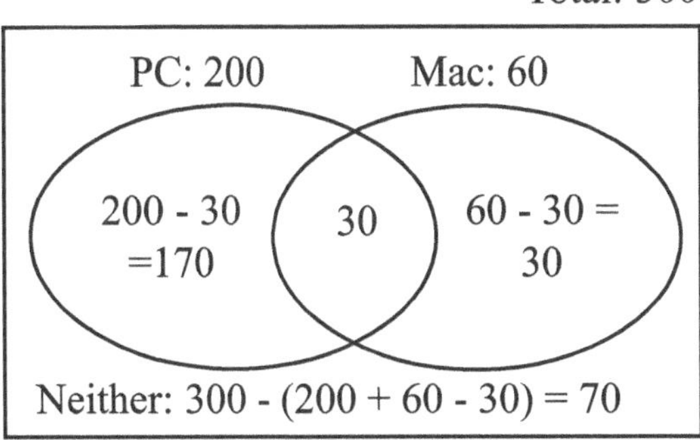

Practice Questions 6.1

1 Of 300 pairs of shorts in a certain clothing store, exactly 160 are casual wear and 200 are swimwear. If 80 percent of the casual wear can be used as swimwear, how many of the pairs of shorts are neither casual wear nor swimwear?

2 At Fantasyland High School, there are 16 members in the math club and 12 members in the chess club. If 10 students belong to only one of the two clubs, how many students belong to both clubs?

3 A store sold 20 gift sets containing pens, pencils, and other stationery. Of these, 12 contained at least one pen, 10 contained at least one pencil, and 4 contained at least one of each. How many sets contained neither a pen nor a pencil?

4 John has 48 customers on his paper route. Of these customers, 10 receive only the Sunday paper, 11 receive the paper only Monday through Saturday, and the remaining customers receive the paper daily. How many customers receive the paper on Sundays?

5 Out of 300 students, 150 were studying Spanish, 80 were studying French, and 50 were studying both languages. How many students were studying neither Spanish nor French?

6.2 :: Fractions in Word Problems

try it yourself Try these two sample questions within 30 seconds.

Q1 Out of a $600 savings account, Tim spent 2/3 of it on clothes and 1/2 of the remainder on books. How much does he have left?

Q2 One monkey ate 2/5 of the apples on a tree and another monkey ate 1/3 of the remaining apples from that tree. If there are 12 apples left on the tree, how many apples were there originally?

general rules

☑ **Forward fraction:** when the original amount is given.

Starting with the original amount, keep subtracting the fractional amounts.

☑ **Backward fraction**: when the original amount is not given.

Let: x = original amount

Equation: $x - xF_1 - xF_2$ = Remaining amount, where F_1 and F_2 are fractions.

F_2 can be $F_3(1 - F_1)$ if it is another part of the remainder,

e.g. $x - xF_1 - xF_3(1 - F_1)$ = Remainder

approach to sample questions

A1 Use a forward fraction:
Remainder after spending on clothes: $600 • 1/3 = $200
Remainder after spending on books: $200 • 1/2 = $100
Therefore, the answer is $100.

A2 Use a backward fraction:
Let: x = # of original apples
Equation: $x - (x)2/5 - (x)1/3 \bullet 3/5 = 12$
Multiply every term by the L.C.D. (15)
$15x - 6x - 3x = 180$
$x(6) = 180$
$x = 30$

Practice Questions 6.2

1 Marcy had 24 stamps, while Jane and Susan had none. Marcy gave 1/2 of her stamps to Jane, and Jane passed 1/3 of her stamps along to Susan. If Marcy then gave Susan 1/3 of her remaining stamps, how many stamps did Susan receive from Marcy and Jane?

2 A ticket agency sold 1/2 of a concert's tickets in the first hour of sale. After 30 more tickets were sold by the second hour, 1/4 of the original number of tickets remained. How many tickets were there to begin with?

3 In a certain company, 3/4 of the workers are male and 2/3 of these males are single. What fraction of the male workers in the company are not single?

4 Every student who studies journalism at a university receives one of the following grades: P, F, W, or C. If 1/6 of the students receive P's, 1/5 receive F's, 1/2 receive W's, and 20 students receive C's, how many students in the university study journalism?

5 Mike ate 1/6 of the total hamburgers in the refrigerator in the morning and then 1/2 of the remainder at lunch. If there were 5 hamburgers left, how many of them were there originally?

6.3 :: Utility Bill / Taxi Fare

try it yourself

Try these two sample questions within 30 seconds.

Q1 The phone company's charge for the first 3 minutes of phone use is \$0.10 and the charge for each additional minute thereafter is \$0.05. If Jean's phone call to her friend on a certain day was 4 hours long, how much must be removed from her allowance this month to cover for this expense?

Q2 An electric company charges \$a for the first 100 kw of electricity used and \$p for each additional kw used. If the total electricity bill for a particular month was \$100, what was the total amount of electricity used in that month?

(A) $\frac{(100 + 100p)}{(p - a)}$ (B) $\frac{100 + 100p - a}{p}$ (C) $\frac{100 + 100p - a}{100p}$

(D) $\frac{100p - a}{p}$ (E) $\frac{100p - a}{100p}$

general rules

☑ The charge for **utility** use such as phone and electricity is based on differential rates for different usage levels: a fixed base charge and an additional proportional charge.

☑ If f = fixed base charge
p = proportional rate per unit of usage after m units of minimum use
T = total usage,
Total charge = f + p(T - m)

approach to sample questions

A1 Total charge $= f + p(T - m)$
$= 0.10 + (.05)(240 - 3)$
$= 0.10 + 11.85$
$= \$11.95$

A2 Let: x = amount of electricity used in that month.
Equation according to the formula: $a + p(x - 100) = 100$
$a + px - 100p = 100$
$px = 100 + 100p - a$

Therefore, $x = \frac{(100 + 100p - a)}{p}$. The answer is (B).

Practice Questions 6.3

1 A telephone call costs \$1.50 for the first 3 minutes and \$0.20 for each minute thereafter. At this rate, what is the cost of a 10-minute telephone call?

2 A long distance phone company's charge for regular hours between two cities is \$1.20 for the first 4 minutes and c cents for each additional minute. The total charge is reduced by 80 percent on calls made after 12:00 p.m. Then, the cost in dollars of a 30-minute call made after 12:00 p.m. between these two cities is

(A) $12 + 4.8c$
(B) $12 + 30c$
(C) $24 + 0.2c$
(D) $24 + 2.6c$
(E) $24 + 5.2c$

3 The rate for party equipment rental is 50 cents for the first hour and 30 cents for each additonal hour. What would the cost be, in dollars, if a rental was for 58 hours?

4 The charge for the first quarter hour of parking is $1.40 and the charge for each additional quarter hour of parking is $1.00. If John paid $7.40 total for parking, for how many hours did he park?

5 A truck-rental company rents trucks at a flat rate of $150 for the first week and $20 for each additional day. In addition, the first 1000 km are free and each kilometer thereafter is $0.50. If a truck was rented for 57 days and the odometer showed that a total of 2,500 km was traveled, what was the amount, in dollars, owed to the rental company?

6.4 :: Cost/Production Comparison

try it yourself

Try these two sample questions within 30 seconds.

Q1 If one apple, which costs 15 cents, is compared to 25 apples, which cost 4 dollars, in terms of price per apple, by how much is the former less expensive than the latter?

Q2 The price of copying was reduced from 9 cents per 2 copies to 15 cents per 4 copies. With $1.80, how many more copies can one make now than before the price reduction?

general rules

☑ **For a fair comparison, change everything into the same unit.**

$$\frac{\text{Total cost (\# workers or machines)}}{\text{Total \# of products}} \quad \text{vs} \quad \frac{\text{Total cost (\# workers or machines)}}{\text{Total \# of products}}$$

Compare.

approach to sample questions

A1 When comapring one apple that costs 15 cents with 25 apples that cost 4 dollars, use a common unit, which is in this case, cents.
The question asks about the discrepancy between the prices per apple.

$$\frac{15 \text{ cents}}{1 \text{ apple}} \quad \text{vs.} \quad \frac{400 \text{ cents}}{25 \text{ apples}}$$

$$\frac{15 \text{ cents}}{1 \text{ apple}} \quad \text{vs.} \quad \frac{16 \text{ cents}}{1 \text{ apple}}$$

Thus, the difference is 1 cent/apple.

The number of copies per cent before the price reduction is 2/9, while the number after it is 4/15.
So, before the price reduction, one could make 180(2/9) = 40 copies, and after the price reduction, one could make 180(4/150) = 48 copies.
Therefore, the difference is 48 - 40 = 8 copies.

Practice Questions 6.4

1 At a fish market, 1 pound of cooked fish costs k dollars. Raw fish, however, costs 4k dollars per 5 pounds. How much, in terms of k, can you save per pound by buying raw rather than cooked fish?

(A) 0.1k
(B) 0.2k
(C) 0.3k
(D) 0.4k
(E) 0.5k

2 A container of 20 cubic meters holds x + 3 items, where x is a positive integer, while a container of 15 cubic meters holds x items. If the two containers are used equally in efficiency, what is the value of x?

(A) 5
(B) 6
(C) 7
(D) 8
(E) 9

3 The price of oranges was initially set at “5 for $4.00.” However, due to a bad harvest, the price was raised to “4 for $5.00.” How much more does it cost per orange now than before?

(A) $0.30
(B) $0.45
(C) $0.50
(D) $0.60
(E) $0.65

CHAPTER 6 TEST

1 At a certain high school, 45 percent of the first-year class takes Spanish and 25 percent takes French. How many first-year students are not enrolled in either class?

(A) 10
(B) 20
(C) 25
(D) 55
(E) Cannot be determined

2 Elizabeth ate 1/4 of a sandwich for breakfast, and then 1/2 of the remainder for dinner. What fraction of the sandwich is left?

(A) 1/6
(B) 1/5
(C) 1/3
(D) 3/8
(E) 2/3

3 John made a phone call that cost $2.60 for the first three minutes and $.40 for each additional minute. If the total charge was $15.00, how many minutes was John on the phone?

4 One month, 300 of a certain product cost x dollars total. The following month, 240 of the same product cost 0.6x dollars total. By what percent did each product drop in value?

(A) 15%
(B) 20%
(C) 25%
(D) 30%
(E) 35%

5 Of 150 seniors in college, 30 are taking history but not English, 80 are taking English but not history, and 10 students are taking neither of the classes. How many students are taking both history and English?

6 Mrs. Dickson spent 40 percent of the money she had in her purse on lunch and 50 percent of what was left after lunch on groceries. If she then had $3 left, how much had she spent on lunch?

(A) $4
(B) $5
(C) $6
(D) $8
(E) $10

7 If the price of a sheet of paper was raised from 5 cents per sheet to 25 cents for 2 sheets, what was the increase in price per sheet?

(A) 2.5 cents
(B) 3 cents
(C) 5 cents
(D) 7.5 cents
(E) 10 cents

8 The cost of renting a truck is $5.00 for the first 2 hours and $1.00 for each additional hour. If the rental extends for more than 10 hours, there is a discount of $0.20 for each hour beyond 10 hours. If the total charge for the rental was $14.60, for how many hours was the truck rented?

Geometry

7.1:: Angle Relations

try it yourself Try these two sample questions within 30 seconds.

Q1 In the figure below, what is the value of x?

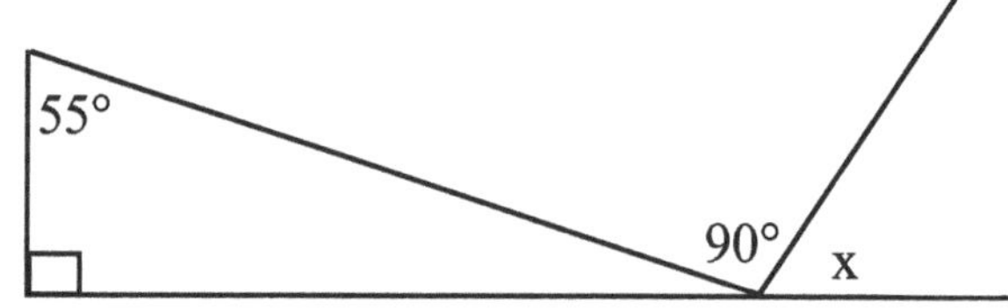

Q2 In the figure below, what is the value of a + b + c + d?

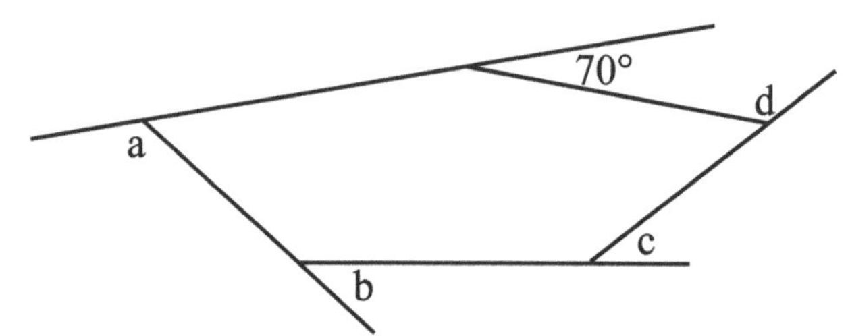

general rules

☑ A **straight line** is 180°.

☑ The sum of the measures of the **interior angles of a triangle** is 180°.

☑ The sum of the measures of the interior angles of a **convex polygon** with n sides is (n - 2)180° or (# of triangles in the polygon) • 180°.

☑ The measure of an exterior angle of a triangle is the sum of the measures of its two remote interior angles.

$m\angle 3 + m\angle 4 = 180°$

$m\angle 1 + m\angle 2 + m\angle 3 = 180°$

$m\angle 4 = m\angle 1 + m\angle 2$

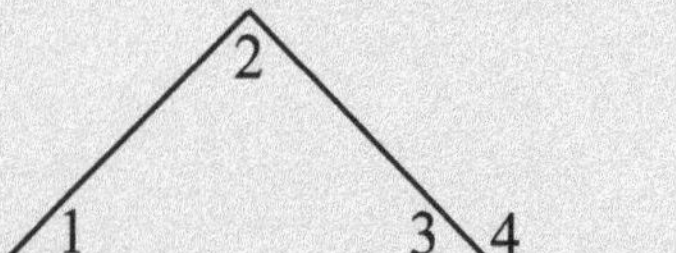

☑ The sum of the measures of the exterior angles of an **n-gon** is 360°.

☑ A line that **bisects** an angle divides the angle exactly in half.

approach to sample questions

A1 The measure of an exterior angle of a triangle is equal to the sum of the measures of its remote interior angles.

Since $x + 90$ is the exterior angle of the triangle,

$x + 90 = 55 + 90$

Therefore, $x = 55°$.

A2 The sum of the measures of the exterior angles of a pentagon is 360°.

Therefore, $a + b + c + d + 70 = 360$

$a + b + c + d = 290°$.

Practice Questions 7.1

1 In the figure below, what is x in terms of y?

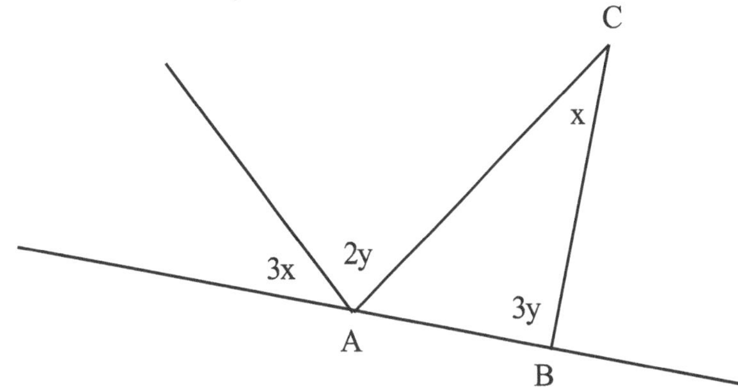

In the triangle below, what is the value of x + y?

(A) 30°
(B) 45°
(C) 60°
(D) 75°
(E) 90°

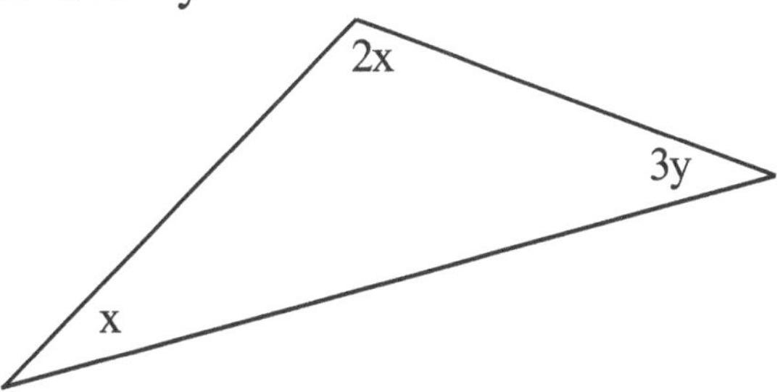

3 A decagon has ten sides. What is the sum of the degree measures of all exterior angles of the decagon?

(A) 90°
(B) 180°
(C) 360°
(D) 1800°
(E) 3600°

4 What is the angle formed by the hour and minute hands of a clock indicating 3:30?

5 In the figure below, if a + b = 110°, find x + y.

(A) 160°
(B) 170°
(C) 180°
(D) 190°
(E) 200°

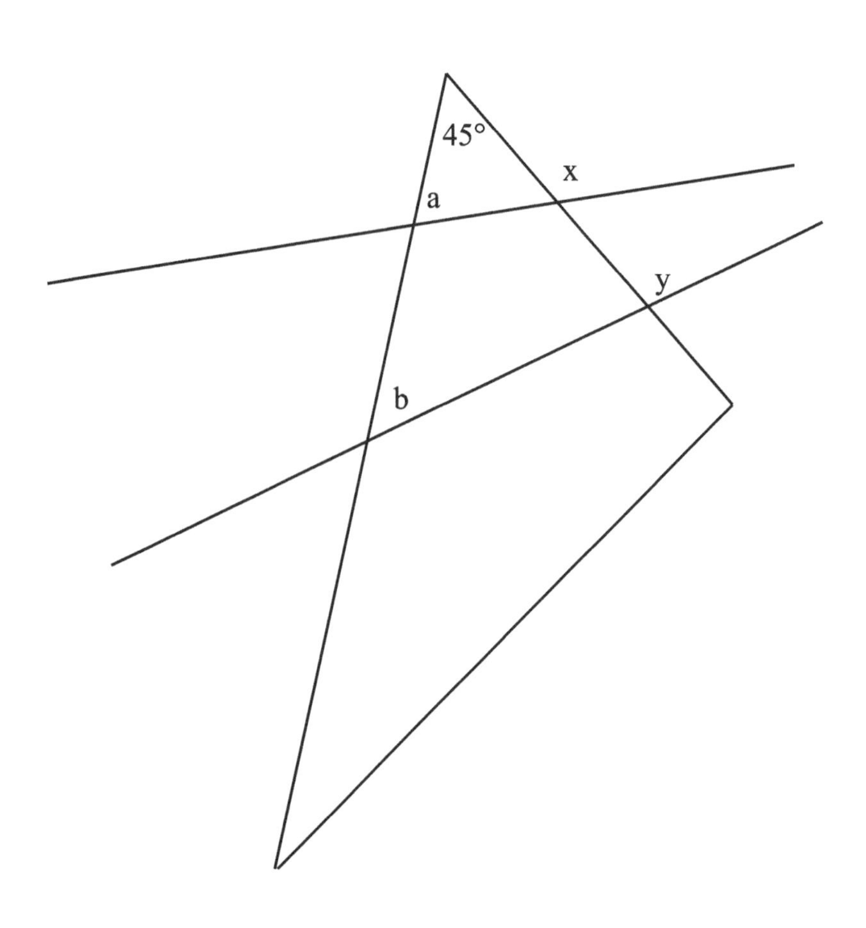

7.2:: Parallel lines

try it yourself Try this sample questions within 30 seconds.

Q1 On the figure below, which of the following statements is/are true?

I. $L_1 // L_2$
II. $M_1 // M_2$
III. $a = 140°$

(A) I only
(B) II only
(C) III only
(D) I and II
(E) None of the above

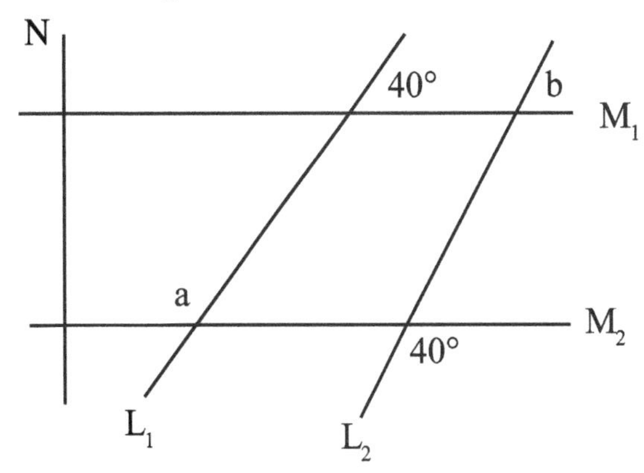

Figure not drawn to scale

TIP: Beware our tendency to read things the way we want to. The fact that lines appear parallel does not necessarily mean that they are, unless otherwise stated.

general rules

☑ Properties of **parallel lines**:

If two parallel lines are cut by a transversal, then their alternate interior angles and corresponding angles are congruent, and their same side interior angles are supplementary.

$m\angle 3 = m\angle 4$
$m\angle 1 = m\angle 4$
$m\angle 1 + m\angle 2 = 180°$

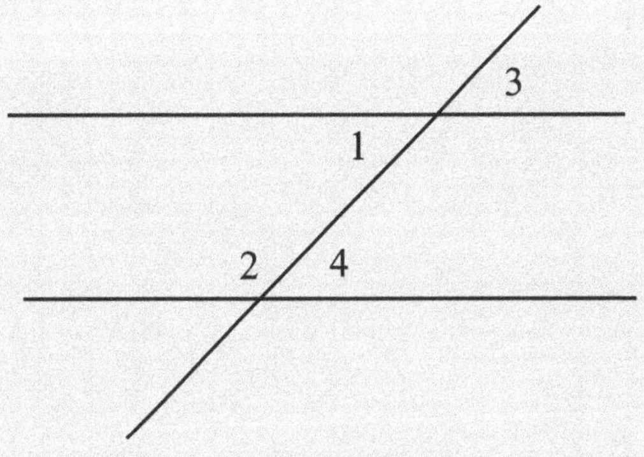

approach to sample questions

A1 Though the picture seems to lure us to believe that the lines are parallel, the given information does not justify it.

In order for the lines to be parallel, b must be 40°.
Though it looks tempting, line N is not perpendicular to neither M_1 nor M_2.
Therefore, the answer is (E).

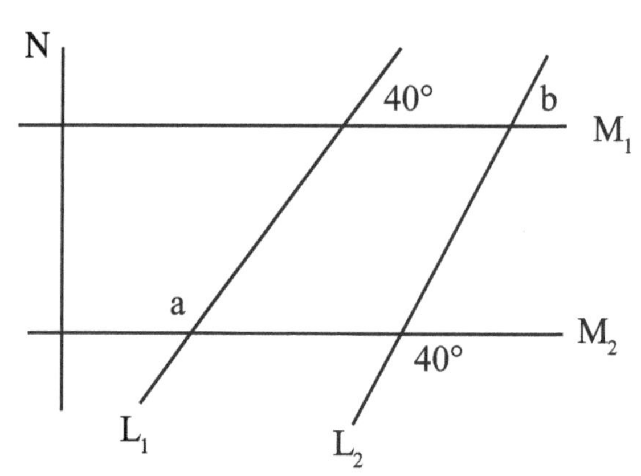

Practice Questions 7.2

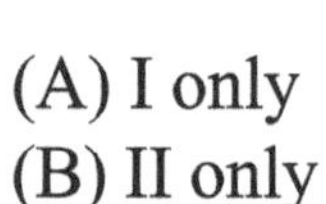

1 In the figure below, if lines L_1 and L_2 are parallel but M_1 and M_2 are not, which of the following cannot be true?

I. $d + f = 180°$
II. $e = g$
III. $b = f$

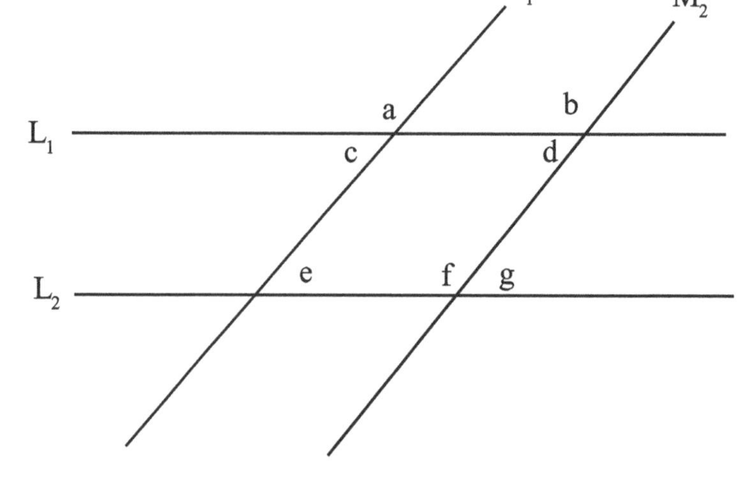

(A) I only
(B) II only
(C) III only
(D) I and II
(E) None

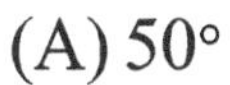

In the figure below, if lines L_1 and L_2 are parallel, rays AC and BC are bisectors of the angles $\angle EAB$ and $\angle ABG$ and $\angle BAD = 130°$, what is the measure of $\angle ACB$?

(A) 50°
(B) 65 °
(C) 75°
(D) 85°
(E) 90°

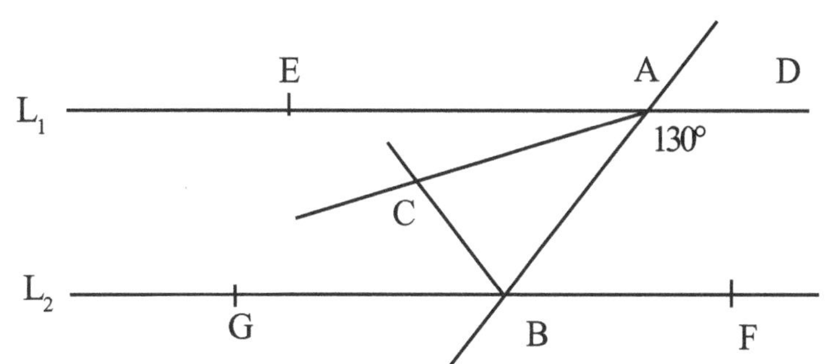

3 In the figure below, if line L_1 and L_2 are parallel and L_3 is a bisector of an angle ABC, what is the value of x in terms of y and z?

(A) y - 2z
(B) y - z
(C) 180 - y + z
(D) 180 - y - z
(E) 2z - y

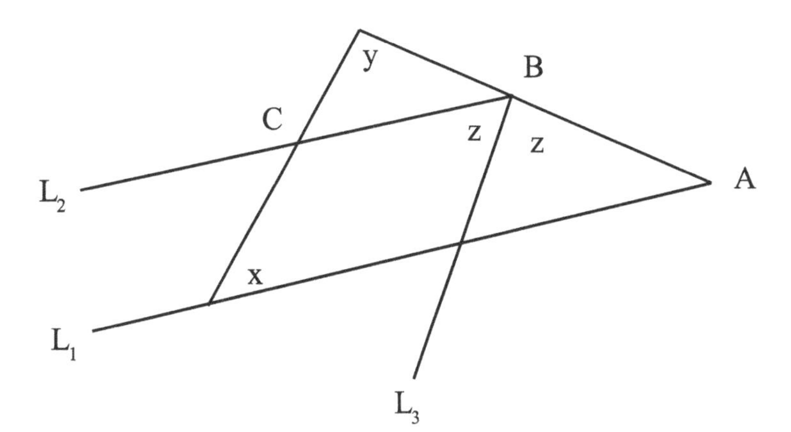

4 In the figure below, if line L_1 and L_2 are not parallel, which of the following could be true?

(A) h = f
(B) b + c = 180
(C) d + a = 180
(D) e = c
(E) d = e

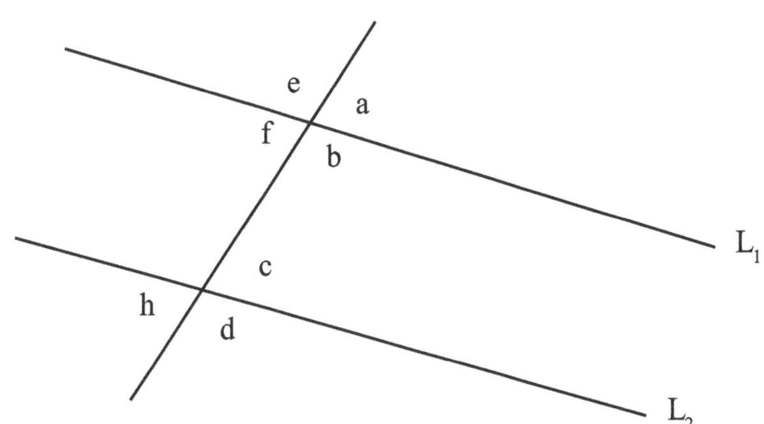

7.3:: Isosceles, Equilateral, & Rt Δ

try it yourself Try this sample question within 30 seconds.

Q1 Which of the following statements could be true?

(A) An altitude drawn from the vertex of an isosceles triangle divides the base into two equal segments.
(B) A right triangle is an isosceles triangle.
(C) An equilateral triangle is always an isosceles triangle.
(D) Isosceles triangles have one angle that is not equal to the other two equal angles.
(E) All statements are true.

general rules

☑ **Isosceles triangle:**
The base angles are equal.

The two sides and two base angles opposite to those sides are equal.

The altitude divides the triangle into two congruent right triangles.

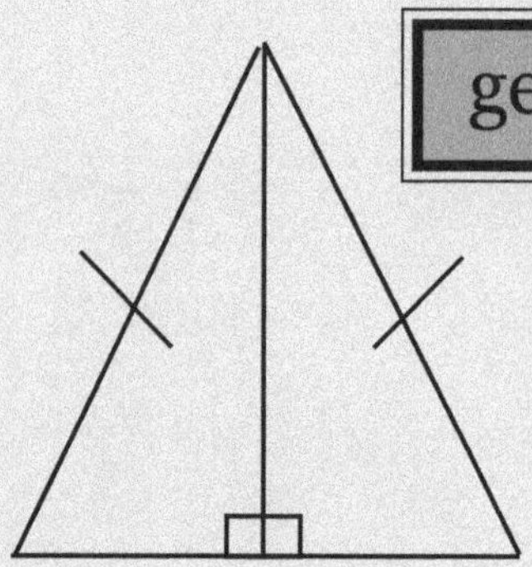

☑ **Equilateral triangle:**
All sides and angles are the same.

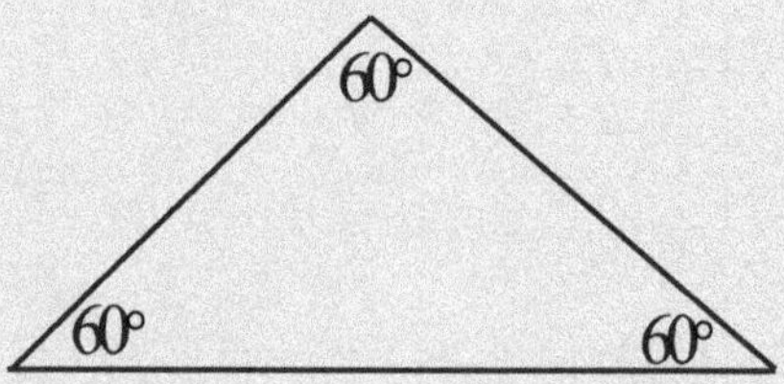

60° 60° 60°

☑ **Right triangle:**
One right angle.

Its two acute angles add up to 90°.

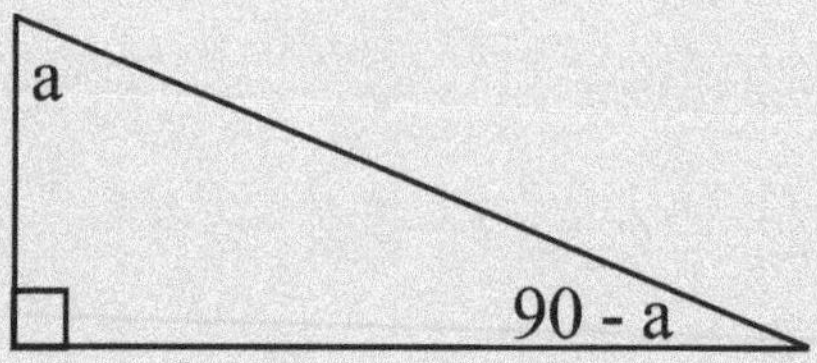

a

90 - a

approach to sample questions

A1 The answer is (C); an equilateral triangle by definition has three equal sides, while an isosceles has at least two equal sides. Therefore, an equilateral triangle is always an isosceles triangle.

(A) is only true if the altitude is drawn from the vertex which joins the two equal sides of the isosceles triangle. The verity of this statement thus depends on from which vertex the altitude is drawn.

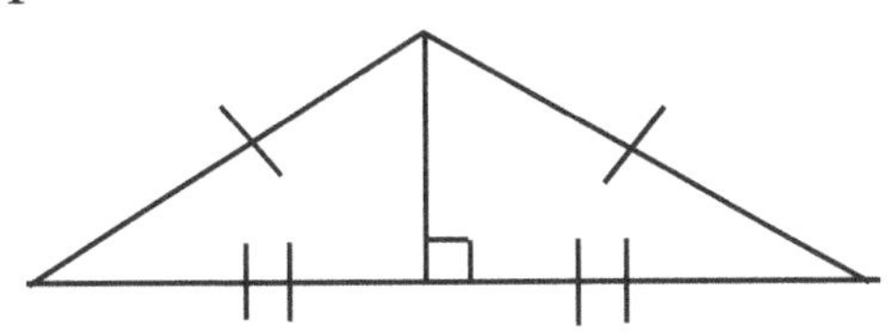

(B) can be true; a right triangle can have two equal legs in which the angles are 45° each (a 45-45-90 triangle). However, a right triangle can also very much have two legs of different length (a 30-60-90 triangle).

(D) can be true in many cases. However, if we are dealing with an isosceles triangle with three equal sides (equilateral triangle), all three angles by definition are equal.

(E) While all statements can be true in certain situations, statement (C) is the only one that holds true regardless of the situation.

Practice Questions 7.3

$m\angle ABD + m\angle BAD =$

(A) 30
(B) 45
(C) 60
(D) 90
(E) 180

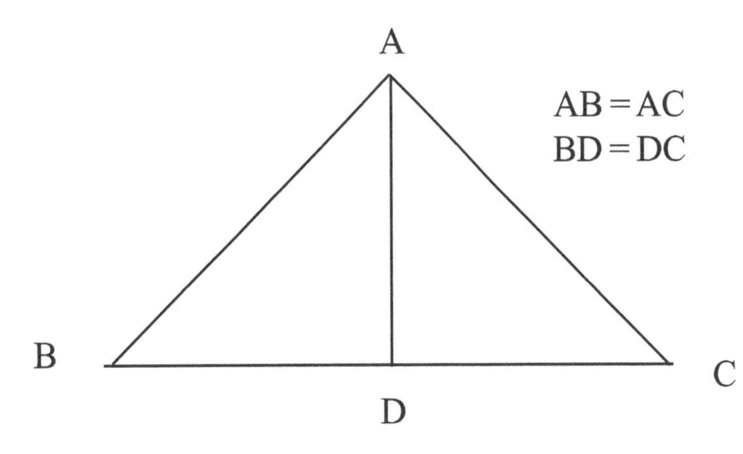

2 In triangle ABC below, which of the following must not be true?

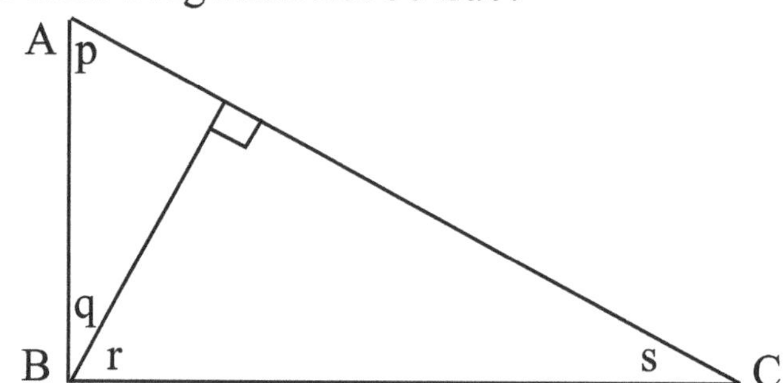

(A) $p + q = 90$
(B) $p + q = r + s$
(C) $p + q + r + s = 180$
(D) $r + s = 90$
(E) $p + s = 90$

3 In the figure below, if m∠BCD is 135 and BA= BC, which of the following must not be true?

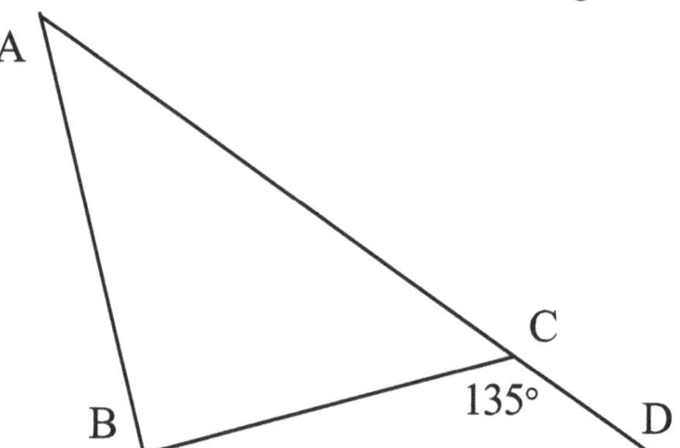

(A) m∠B and m∠C are equal
(B) m∠B = 45
(C) ABC is a right triangle
(D) ABC is an isosceles triangle
(E) ABC is an equilateral triangle

4 In the right triangle below, if x is 30 less than two times y, what is x?

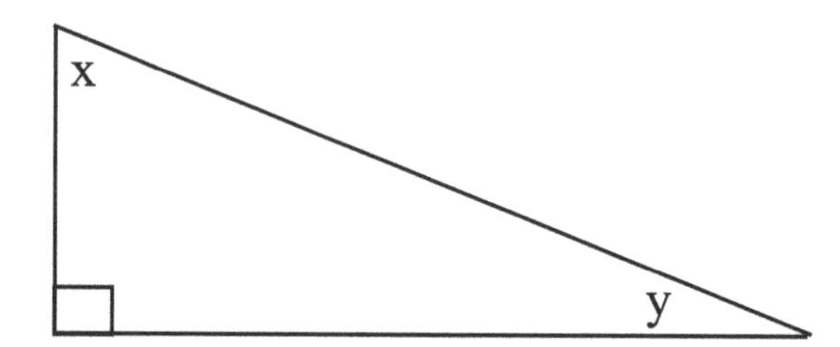

(A) 30
(B) 40
(C) 45
(D) 50
(E) 60

5 In triangle ABC, x =

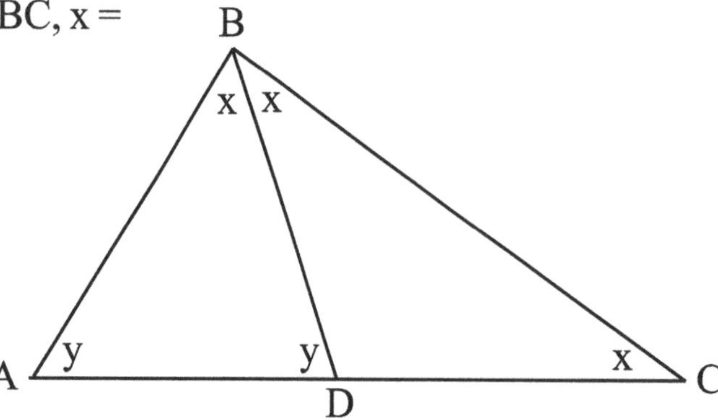

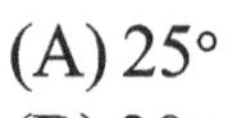

(A) 25°
(B) 30°
(C) 36°
(D) 44°
(E) 45°

7.4:: Special Rt Δ / Pythagorean Thm

try it yourself Try these two sample questions within 30 seconds.

Q1 & Q2

Find x & y.

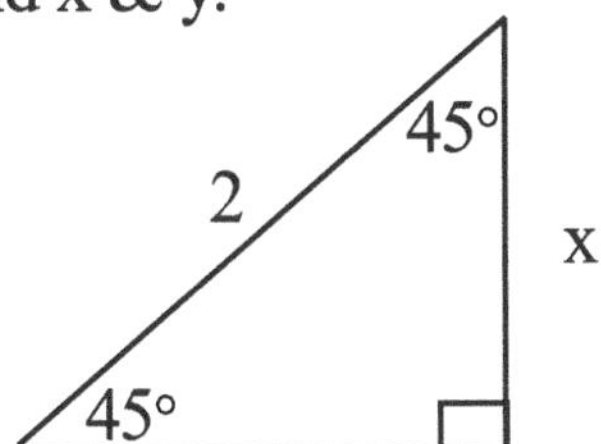

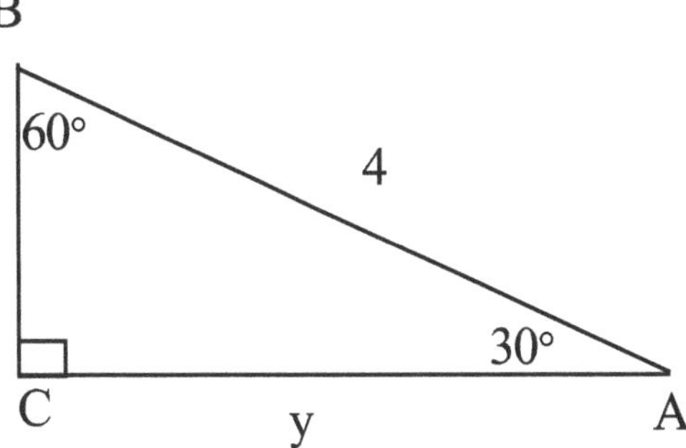

general rules

☑ **Pythagorean theorem in a right triangle**

$a^2 + b^2 = c^2$

(If $a^2 + b^2 < c^2$, then $\angle C$ is an obtuse angle)

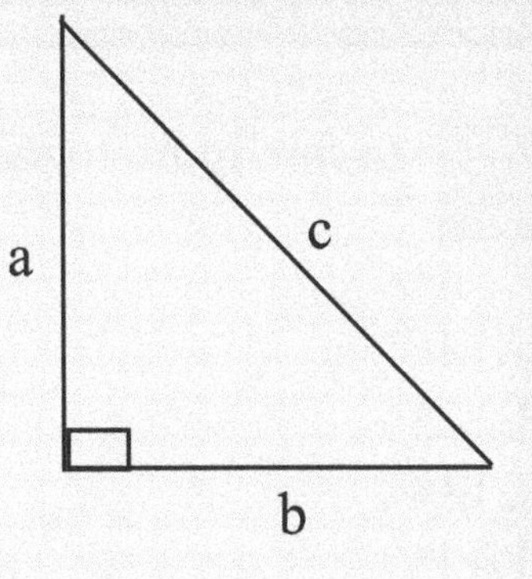

Frequently-used **Pythagorean triples**:

3 : 4 : 5 and their multiples (for example, 6 : 8 : 10)
5 : 12 : 13 and their multiples
7 : 24 : 25 and their multiples
8 : 15 : 17 and their multiples
9 : 40 : 41 and their multiples

☑ **Special right triangles**

45 - 45 - 90 triangle

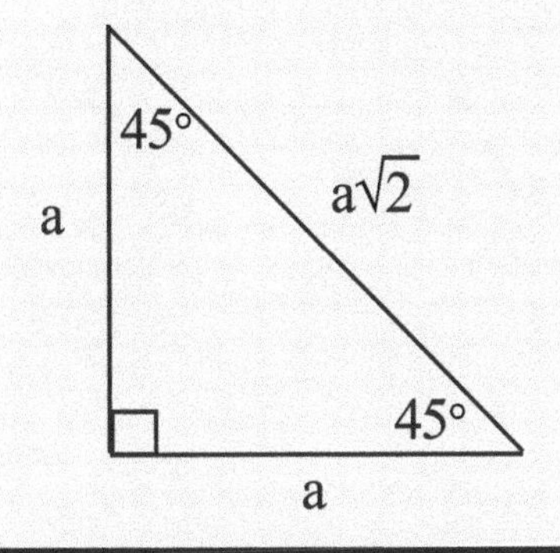

30 - 60 - 90 triangle

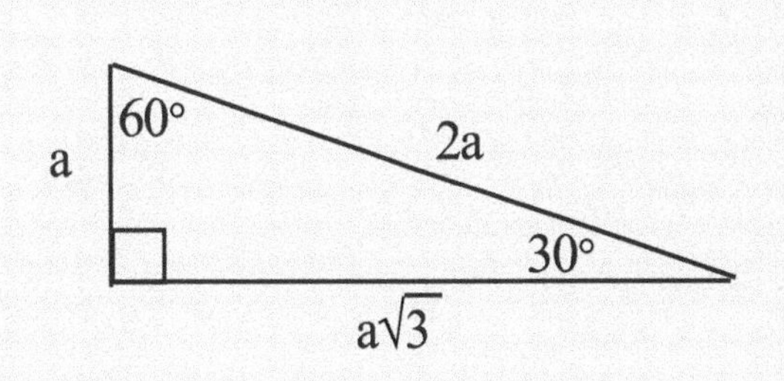

approach to sample questions

A1 The first triangle is a 45-45-90 right triangle, which means that both sides have the length of x and the hypotenuse has the length of $x\sqrt{2}$. Knowing this, we can write

$x\sqrt{2} = 2$; and $x = \frac{2}{\sqrt{2}}$. Rationalizing this, we get $x = \sqrt{2}$.

A2 The second triangle is a 30-60-90 right triangle, which means that the shortest side (corresponding to angle A, which is 30°) has the length x, the longer side (corresponding to angle B, which is 60°) has the length $x\sqrt{3}$, and the hypotenuse has the length 2x. Knowing this,

Since AB = 4, BC = (AB)/2 = 2.
Then, $y = (BC)\sqrt{3} = 2\sqrt{3}$

Practice Questions 7.4

1 In the figure to the right, x =

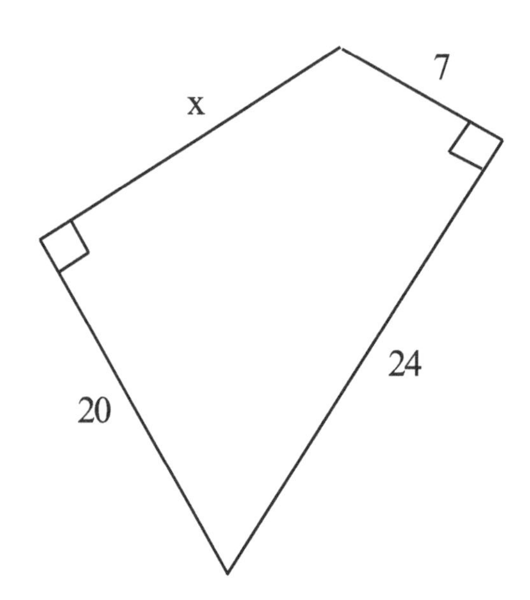

(A) 10
(B) 13
(C) 14
(D) 15
(E) 18

2 In the figure to the right, what is x?

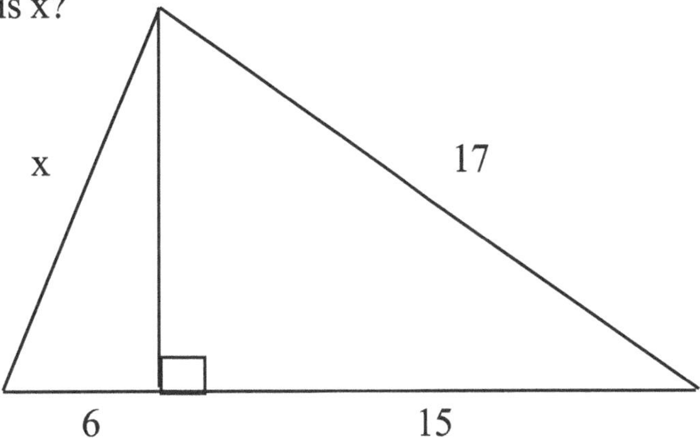

3 2/3 of the length of a rectangle equals 4/3 the width. What is the diagonal of the rectangle in terms of width w?

(A) $w\sqrt{3}$
(B) 2w
(C) $w\sqrt{5}$
(D) $w\sqrt{6}$
(E) 3w

4 In the figure to the right, what is x?

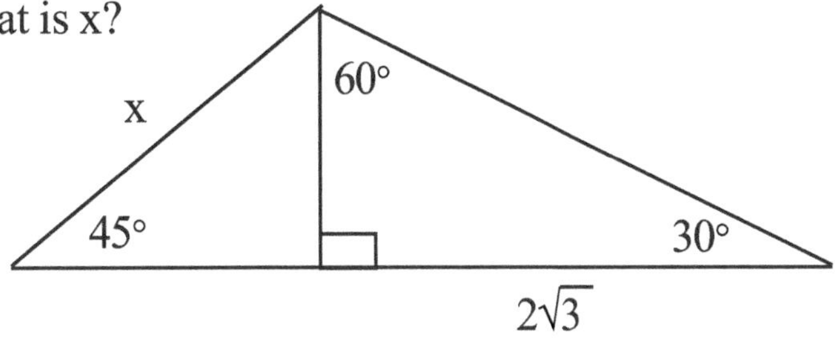

5 If the area of the right triangle below is $\frac{\sqrt{3}}{2}$, what is the value of angle m∠A?

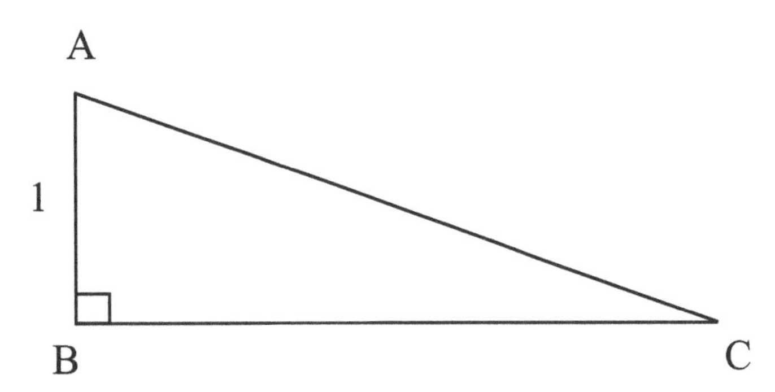

7.5:: Inequality in Triangles

try it yourself

Try this sample question within 30 seconds.

Q1 Tom has an emotional distance of 6 with his father and 16 with his mother. If the emotional distance between his parents is a prime number, how many different distances between the two parents are possible?

(A) 1
(B) 2
(C) 3
(D) 4
(E) 5

general rules

☑ If the length of two sides of a triangle is a and b, then the third side is between $a + b$ and $a - b$, assuming $a > b$.

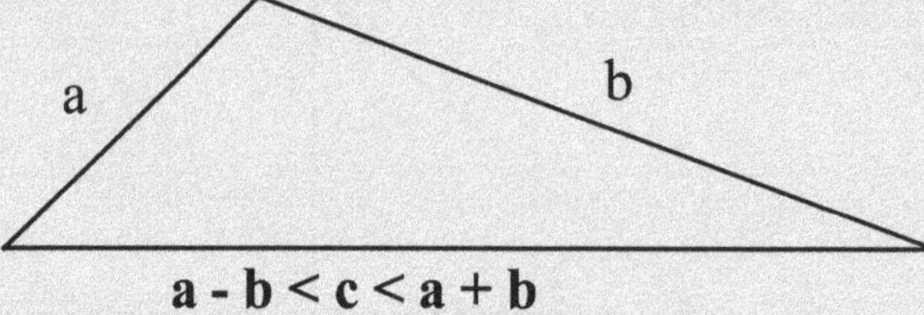

$a - b < c < a + b$

☑ The side of a triangle is greater when its opposite angle is larger.

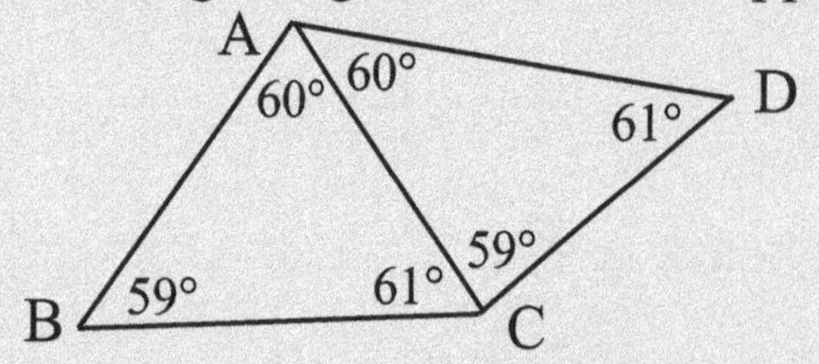

AB > AC > AD

☑ When two adjacent sides of a triangle are the same between two triangles, the third side is greater if its opposite included angle is larger *(Scissors theorem)*

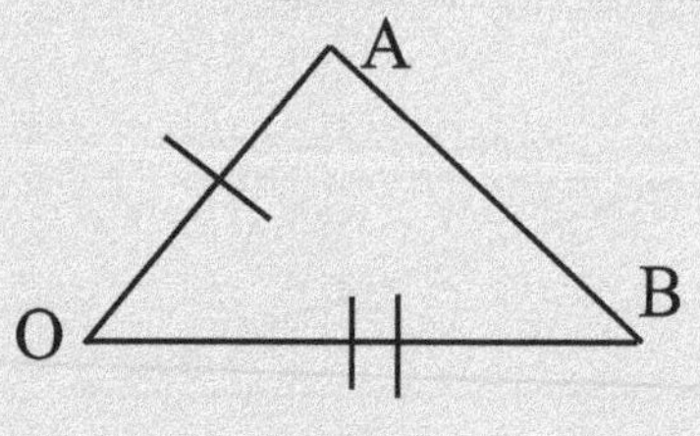

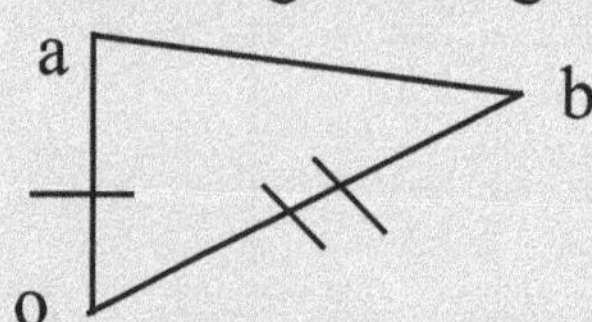

If $m\angle AOB > m\angle aob$, then $AB > ab$

approach to sample questions

A1 By drawing a picture, you could see that one could depict Tom's emotional distance between his parents through a triangle.

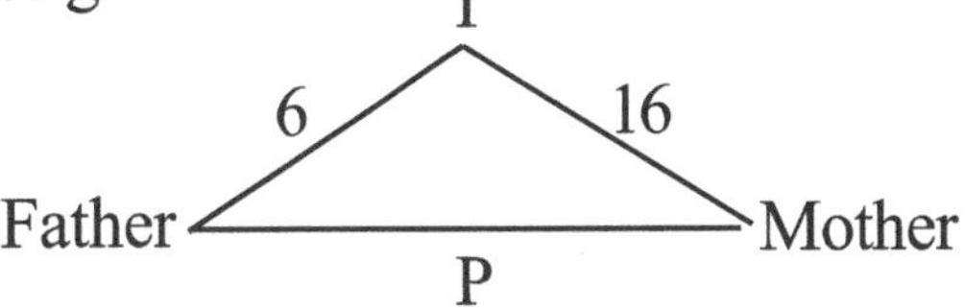

According to the inequality rule of the triangle, the third side, P, is greater than the difference between and less than the sum of the two sides.
$16 - 6 < P < 16 + 6$; thus $10 < P < 22$.

The possible prime numbers between 10 and 22 are 11, 13, 17, and 19. Therefore, there are four possible emotional distances between Tom's parents. The answer is (D).

Practice Questions 7.5

1 Which of the following cannot be the length of side BC?

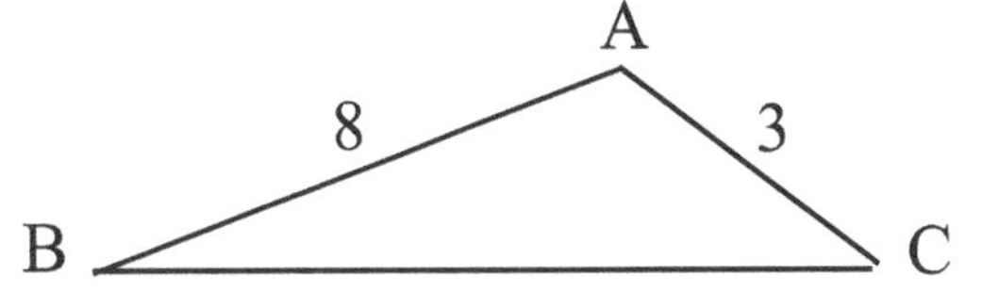

(A) 6
(B) 8
(C) 9
(D) 10
(E) 11

2 Triangle ABC has sides of length 5, 17, and x, where x is the length of the longest side. If x is a prime number, the perimeter of ABC is equal to which of the following?

(A) 41
(B) 45
(C) 51
(D) 53
(E) 59

For the two triangles below, which statement is true?

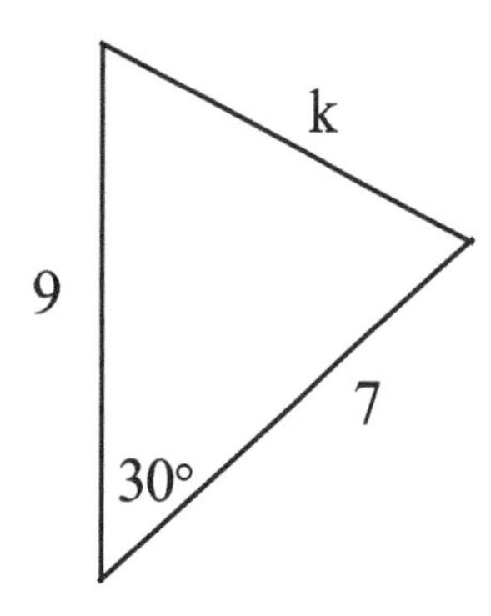

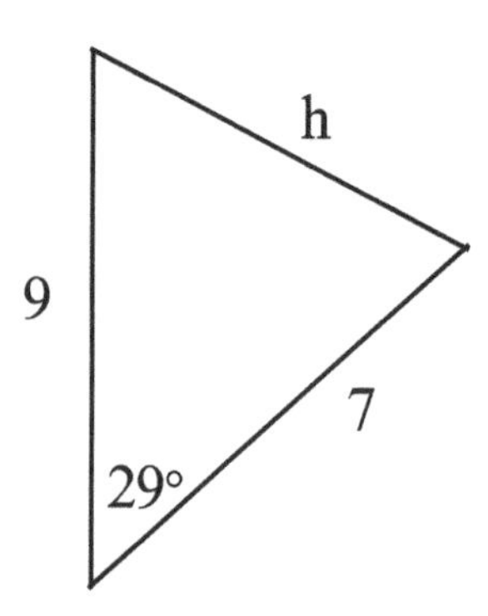

(A) $k = h$
(B) $k > h$
(C) $k < h$
(D) They are both right triangles
(E) None of the above

4 The sides of a triangle are t, $t + 2$, and $t - 1$. Which of the following is the possible measure for the angle opposite to $t + 2$?

(A) 57°
(B) 58°
(C) 59°
(D) 60°
(E) 70°

7.6:: Perimeter & Circumference

try it yourself Try this sample question within 30 seconds.

Q1 John built a circular fence around his house so that no one can trespass on his property. Mary built her own circular fence for her own protection. If John's fence has a radius of 49 feet and Mary's fence has a radius of 35 feet, what is the difference of the lengths of the fences, in feet? (Use 22/7 as π.)

(A) 44
(B) 66
(C) 88
(D) 110
(E) 77

general rules

Perimeters of polygons:

☑ Square

s

$P = 4s$

☑ Parallelogram

$P = 2L + 2W$

☑ Circle

r

$C = 2\pi r$
$= \pi D$

approach to sample questions

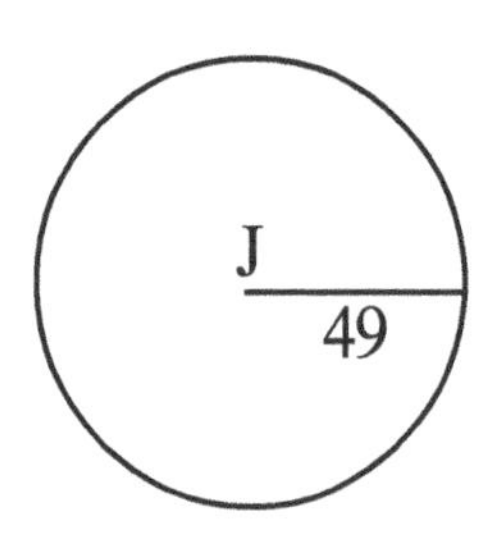

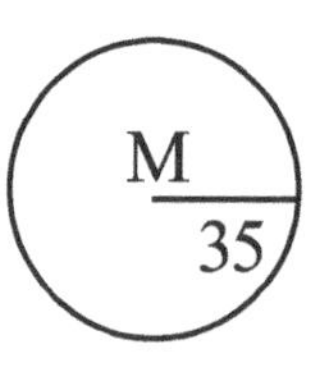

The question is asking about the lengths of these circular fences, which translates into the circumferences of these circles.

John: $2\pi r = 2\ (22/7)(49) = 308$
Mary: $2\pi r = 2(22/7)(35) = 220$

308 - 220 = 88 feet. The answer is (C).

Practice Questions 7.6

1 What is the circumference of a circle with diameter s?

(A) $\pi s/2$
(B) πs
(C) $2\pi s$
(D) $4\pi s$
(E) $4\pi s^3$

2 If the perimeter of the rectangle below is 60, what is the length 2x - 3?

2x - 3

x + 3

(A) 10
(B) 13
(C) 17
(D) 20
(E) 30

3 Compare the perimeters of the following two figures.

t | Fig. 1 | s | t | Fig. 2

(A) $P_1 > P_2$
(B) $P_1 < P_2$
(C) $P_1 = P_2$
(D) Cannot be determined

A wheel has a radius of 14 inches. How many revolutions would it take to wind up a 100 foot long cord? (Use 22/7 for π).

(A) 10
(B) 11
(C) 14
(D) 22
(E) 36

If the perimeter of a rectangle is 3 times its length, then how many times is it of its width?

(A) 4 times
(B) 5 times
(C) 6 times
(D) 8 times
(E) 10 times

The rectangle below depicts a brick-fenced area for a flower bed. If an additional 20 square meters is to be enclosed by moving just one side of the wall in 1 meter increments while maintaining the area's rectangular shape, what is the least possible number of meters of additonal bricks needed?

(A) 0.5m
(B) 1m
(C) 2m
(D) 3m
(E) 4m

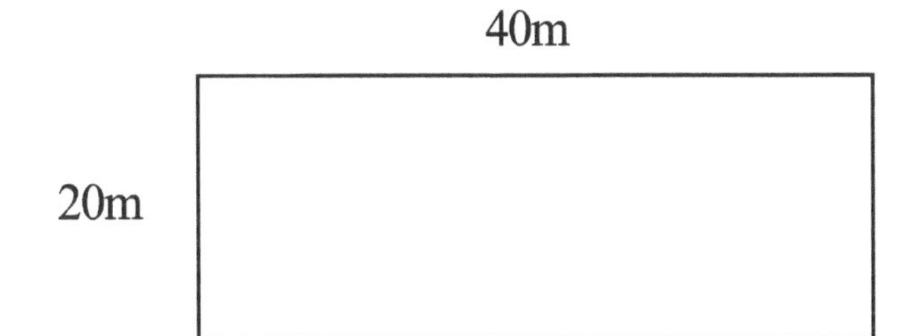

7.7:: Polygons: Features/Area

try it yourself Try these two sample questions within 30 seconds.

Q1 In the right triangle ABC below, what is the altitude drawn to the hypotenuse AC when AB = 5 and BC = 12?

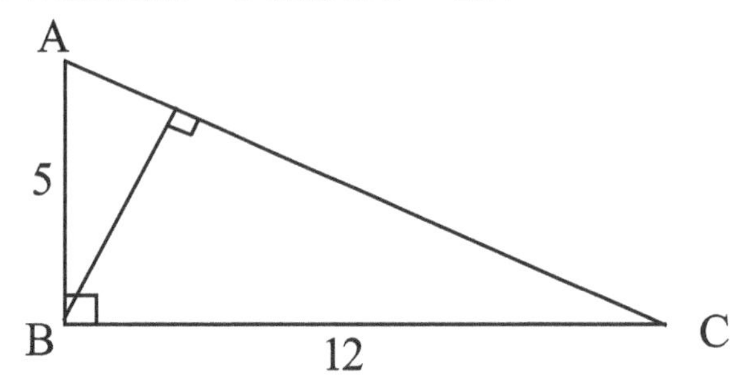

(A) 14/4
(B) 13/3
(C) 69/15
(D) 60/13
(E) 39/8

Q2 If ABCD is a parallelogram, what is the area of ABCD?

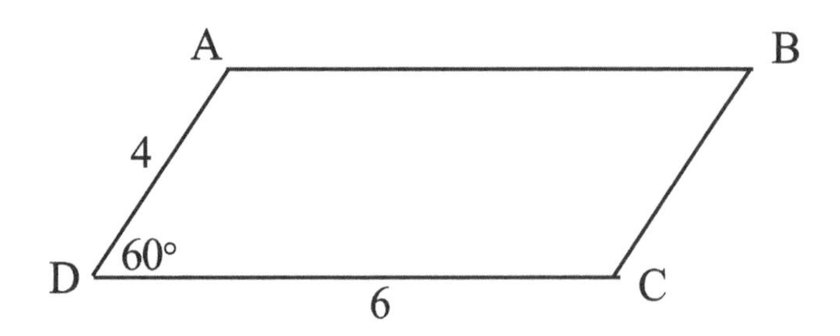

Features/area of polygons

general rules

square

$A = s^2$

rectangle

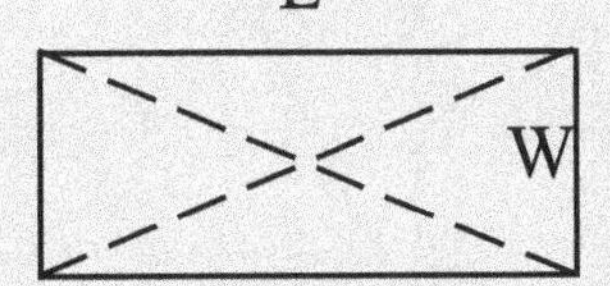

diagonals are equal.

$A = LW$

triangle

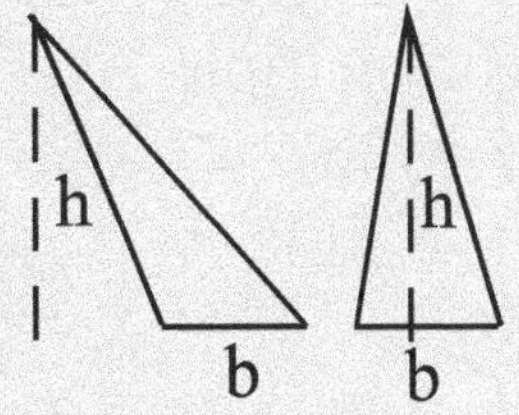

$A = (1/2)bh$

equilateral triangle

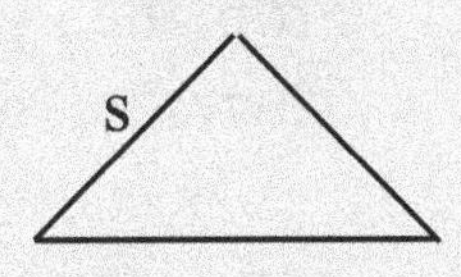

$A = \frac{(s^2)\sqrt{3}}{4}$

right triangle (2 area formulas)

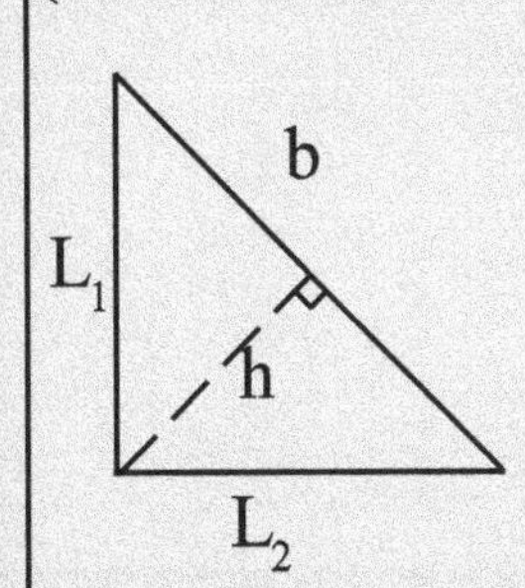

$A = (1/2)(L_1 \bullet L_2)$
$= (1/2)bh$

trapezoid

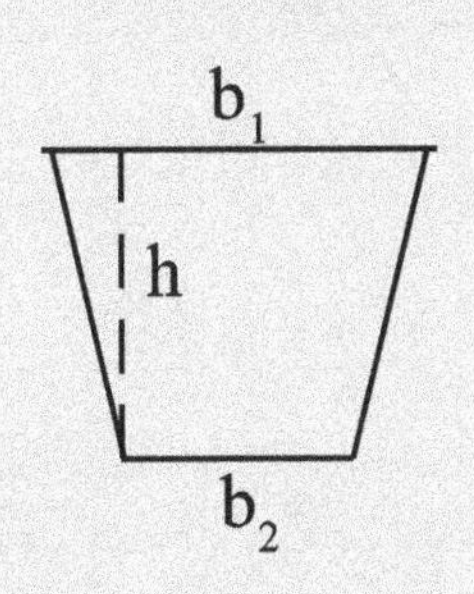

$A = (1/2)(b_1 + b_2)h$

parallelogram

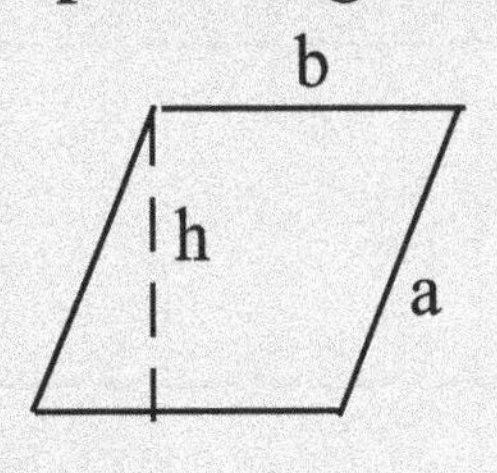

2 pairs of // lines & supplementary adjacent angle pairs

$A = bh$
$A \neq ab$

rhombus

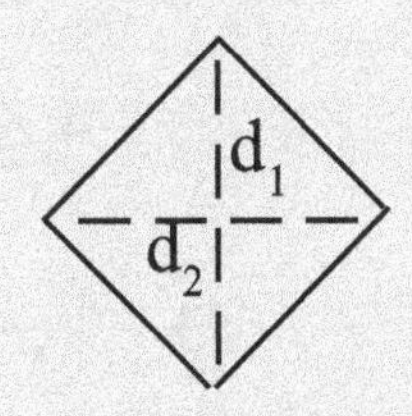

four sides are same & diagonals (d_1, d_2) are perpendicular

$A = d_1 \bullet d_2$

circle

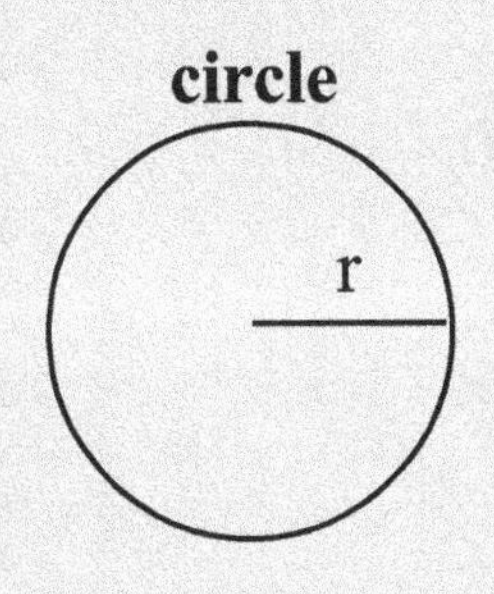

$A = \pi r^2$

An **Isosceles trapezoid** has two equal legs, base angles and diagonals

The area of a **parallelogram** is less than the product of the two adjacent sides. $bh < ba$

A **kite** has perpendicular diagnoals but unequal sides

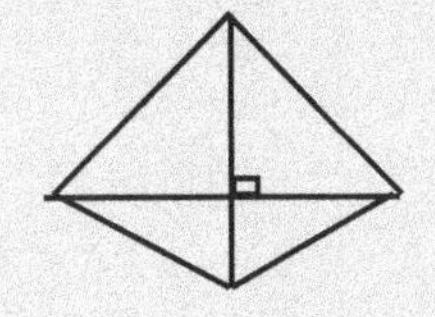

approach to sample questions

A1 There are two different ways to obtain the area of a right triangle:

Let: h = altitude

(1/2)(AB • BC) = (1/2)(AC • h)

From the Pythagorean triple, AC = 13.

Therefore, 5(12) = 13h

h = 60/13.

The answer is (D).

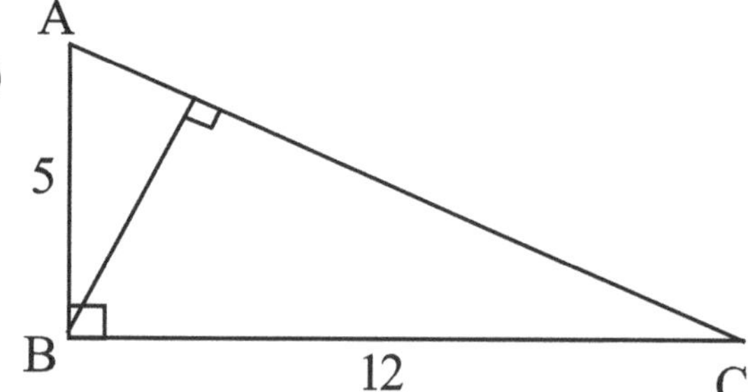

A2 Drawing an altitude from angle A forms a 30-60-90 triangle, which has a height of $2\sqrt{3}$.

Area ABCD = bh = $6(2\sqrt{3}) = 12\sqrt{3}$.

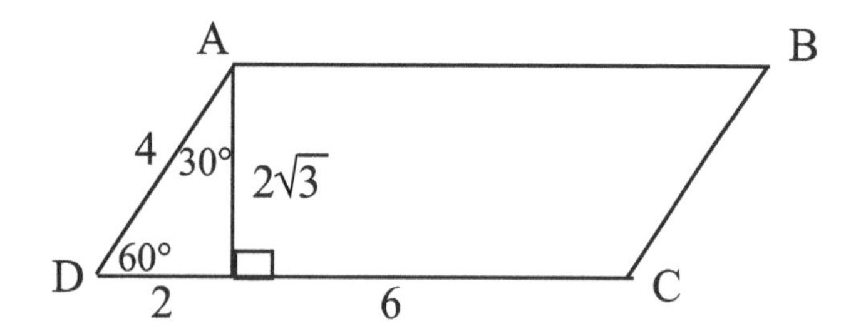

Practice Questions 7.7

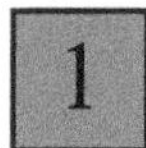

1 What is the area of ΔABC?

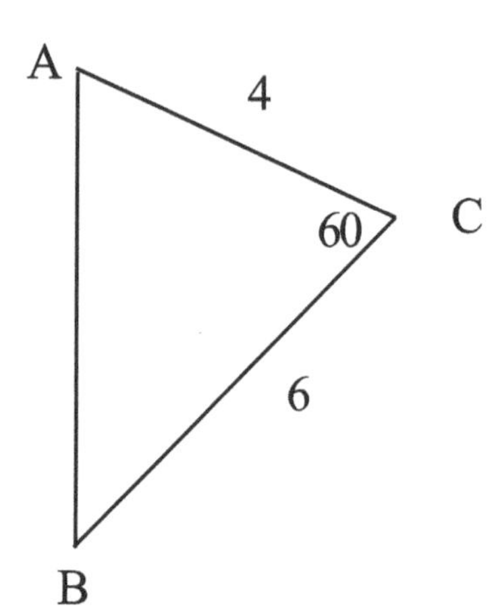

(A) $4\sqrt{3}$
(B) 5
(C) $5\sqrt{3}$
(D) 6
(E) $6\sqrt{3}$

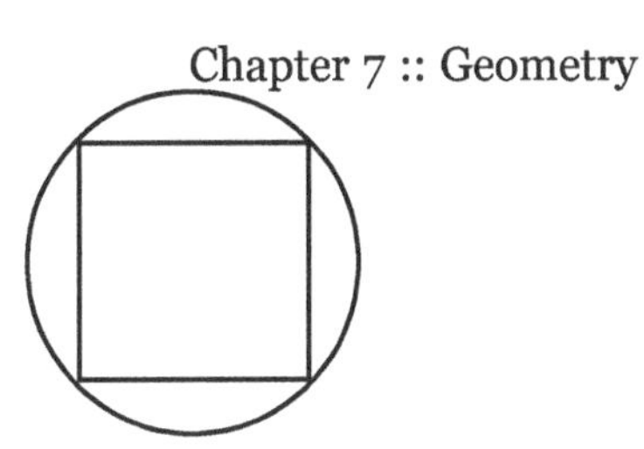

2 In the figure to the right, if a square is inscribed in a circle with area 4π, find the area of the square.

(A) 6
(B) 8
(C) $4\pi - 4$
(D) 3π
(E) 12

3 In the three-dimensional figures below, semicircle AC of radius 1 shares one side with equilateral triangle ABC. What is the area of ΔABC?

(A) $\sqrt{3}$
(B) 1
(C) 1/2
(D) 2π
(E) $\pi/2$

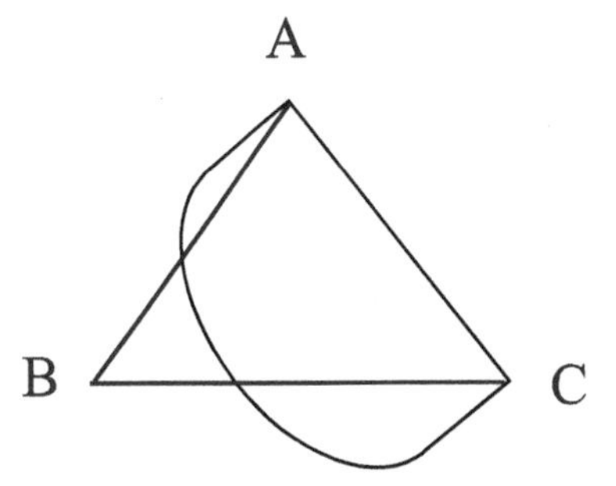

4 An isosceles right triangle has area 32. What is the length of the hypotenuse of the triangle?.

(A) $6\sqrt{3}$
(B) $8\sqrt{2}$
(C) $8\sqrt{3}$
(D) 16
(E) $16\sqrt{2}$

7.8:: Bird's-Eye View in A/P

try it yourself Try this sample question within 30 seconds.

Q1 Figures A and B are composed of five same-sized squares. Which statement is true about the perimeters of the two figures?

(A) $P_A > P_B$
(B) $P_A < P_B$
(C) $P_A = P_B$
(D) Cannot be determined

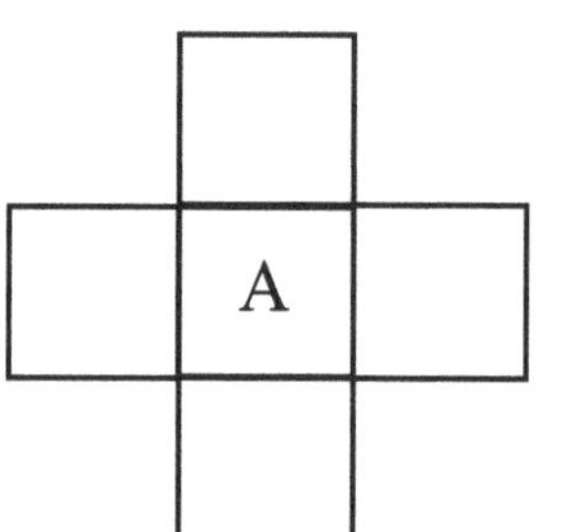

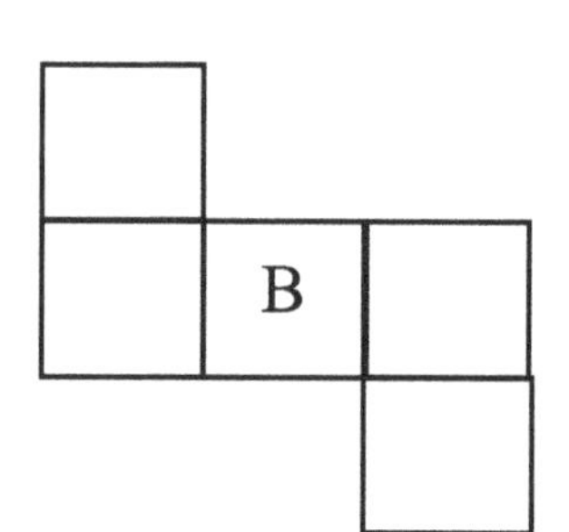

general rules

Bird's-Eye View in perimeter problems:
To a bird flying in the sky, the perimeters, or lengths of the three figures below are equal.

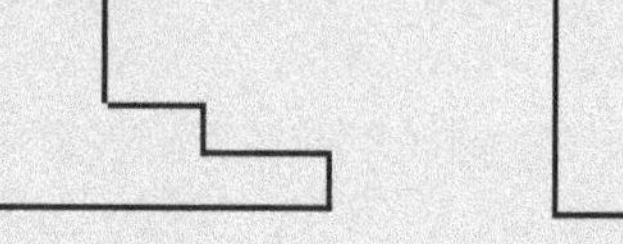

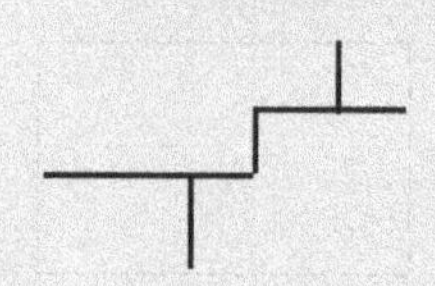

Bird's-Eye View in area problems

☑ **Matching technique:** we can bring separated parts together to form a whole figure

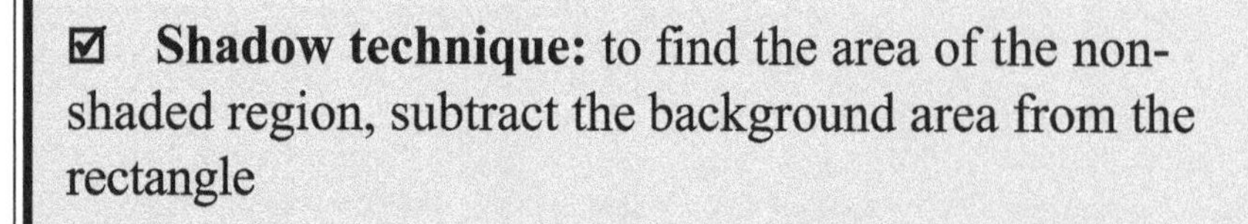

☑ **Shadow technique:** to find the area of the non-shaded region, subtract the background area from the rectangle

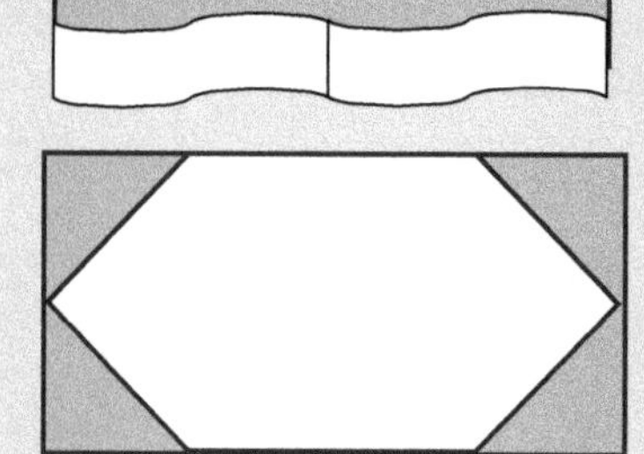

☑ **Different perspective**: we can approach the same problem from a different angle to get the same answer.
For example, the area for a right triangle is either (1/2)bh or (1/2)($L_1 \bullet L_2$) depending on which way the triangle is viewed.

approach to sample questions

A1 Using the Bird's-Eye View method, we can figure out that the perimeters of the two figures are equal. The answer is (C).

Practice Questions 7.8

1

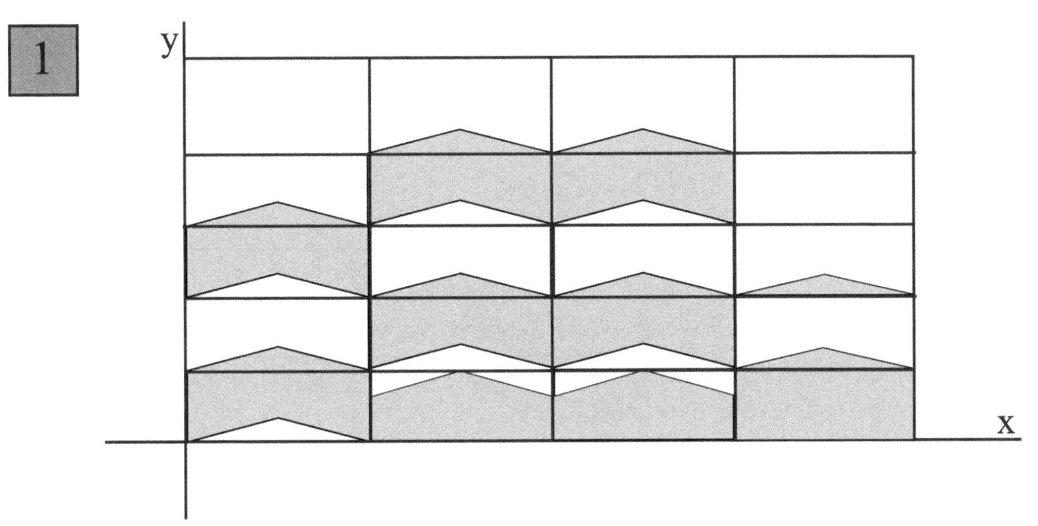

In the above figure, the boundary of the shaded region is made up of line segments and one block of the area is 1 square unit. What is the area of the shaded region?

(A) 8 (B) 8.2 (C) 8.5 (D) 9 (E) 9.1

2 If all line segments in the figure below are either completely vertical or horizontal, which statement is true?

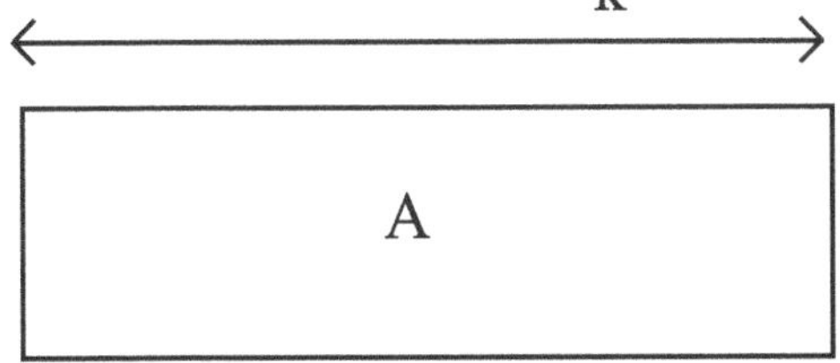

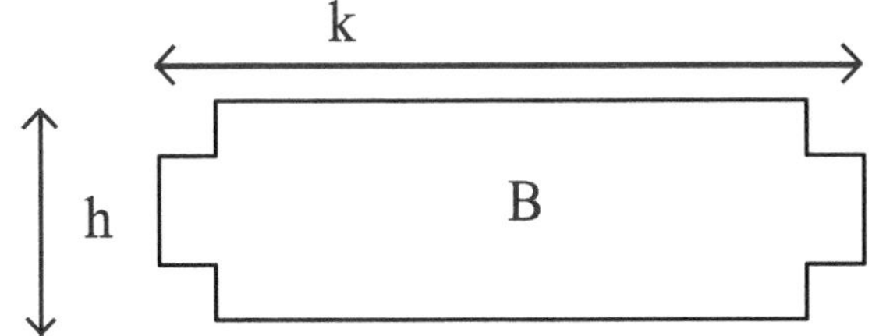

(A) perimeter A < perimeter B
(B) perimeter A = perimeter B
(C) perimeter A > perimeter B
(D) area A = area B
(E) area B = hk

3 In the figure below, all corners are right angles. What is the perimeter?

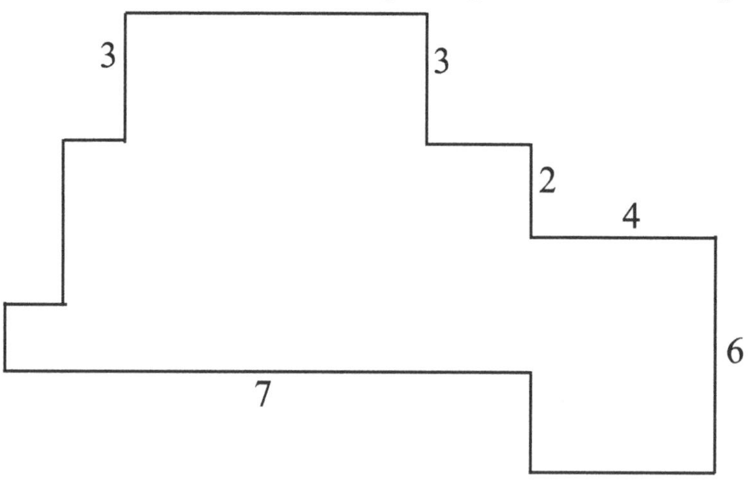

(A) 38 (B) 42 (C) 44 (D) 48 (E) 52

4 Find the area of the shaded triangle in the rectangle to the right.

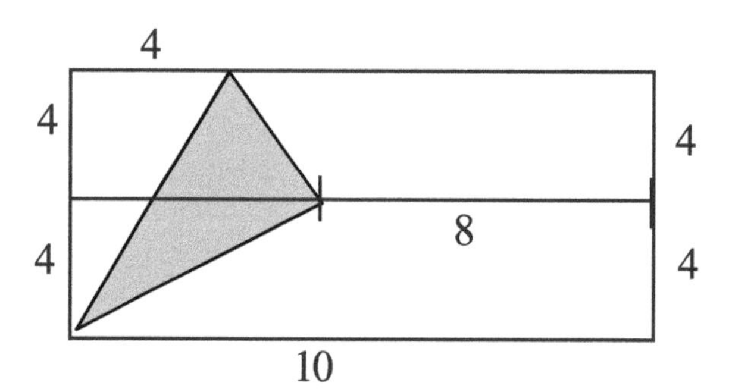

(A) 9 (B) 10 (C) 16 (D) 20 (E) 25

5 The area of the equilateral triangle below is 16 and D, E, F, and G are the mid-points of segments AB, BC, AC, and DE, respectively. Find the area of triangle DFG.

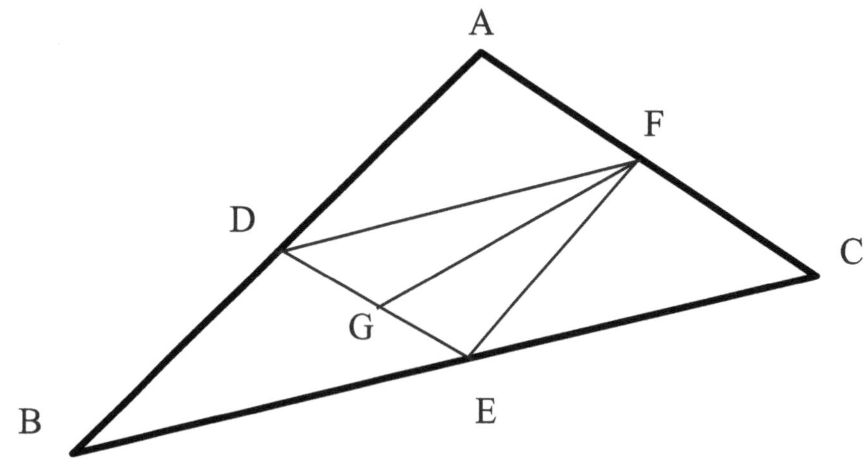

(A) 1 (B) 2 (C) $2\sqrt{2}$ (D) $4\sqrt{3}$ (E) 4

7.9:: Circles & Related Topics

try it yourself

Try these three sample questions within 60 seconds.

Q1 The circle with center O is divided into 8 equal areas. If the entire area of the circle is 16π, what is the perimeter of the area A?

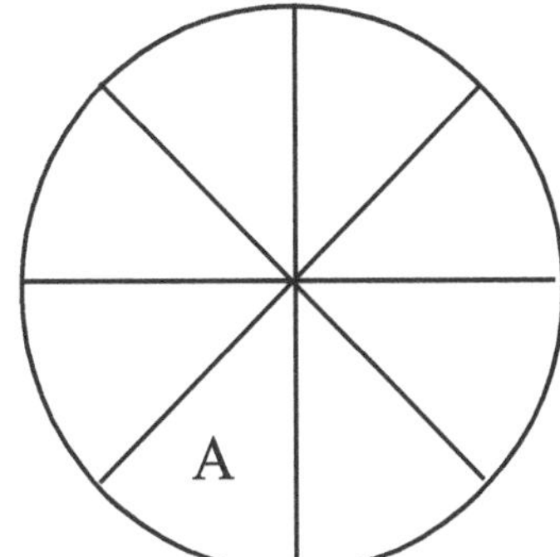

(A) 8
(B) $8 + \pi$
(C) $8 + 2\pi$
(D) $16 + \pi$
(E) $16 + 2\pi$

Q2 In the figures below, two circles with radius 2 and 1 respectively have sectors X and Y, whose areas are equal. If the central angle AOB is 35°, what is the central angle COD?

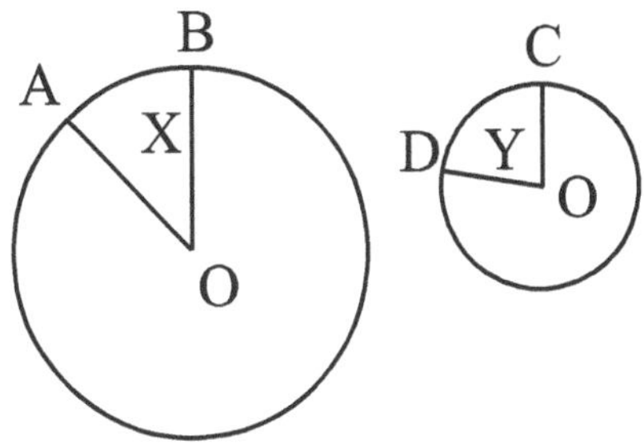

(A) 35°
(B) 55°
(C) 70°
(D) 90°
(E) 120°

Q3 If a 6 by 8 rectangle is inscribed in a circle, what is the area of the circle?

(A) 9π
(B) 10π
(C) 16π
(D) 25π
(E) 50π

Arc & part-of-a-circle problems

general rules

☑ Length of an arc = $\frac{m\angle \text{central angle} \times}{360}$ (circumference)

☑ Area of a part of a circle = $\frac{m\angle \text{central angle} \times}{360}$ (area of circle)

Intersection between different circles

☑ Common chords and radii are perpendicular to each other.

Polygon inscribed/circumscribed around circle

☑ A triangle with the **vertex on the center** is always an **isosceles** triangle.

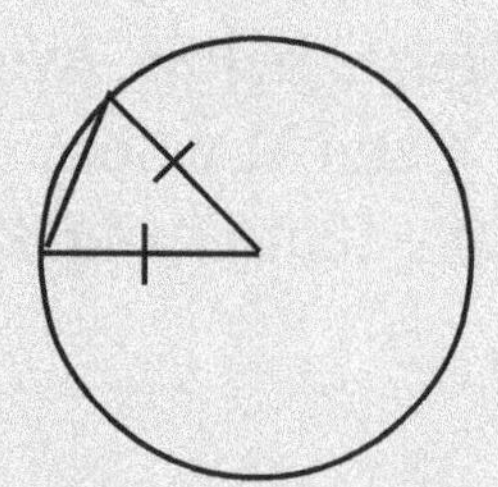

☑ A triangle inscribed with the **diameter as one side** is always a **right** triangle.

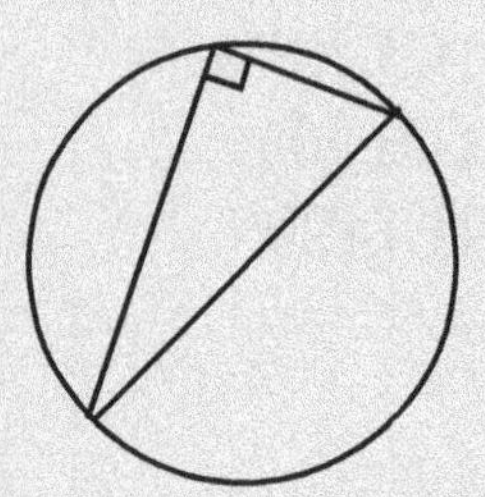

☑ An angle formed between the **radius and a tangent** line is always a **right** angle.

The length of two exterior tangent lines drawn from an outside point to a circle are the same.

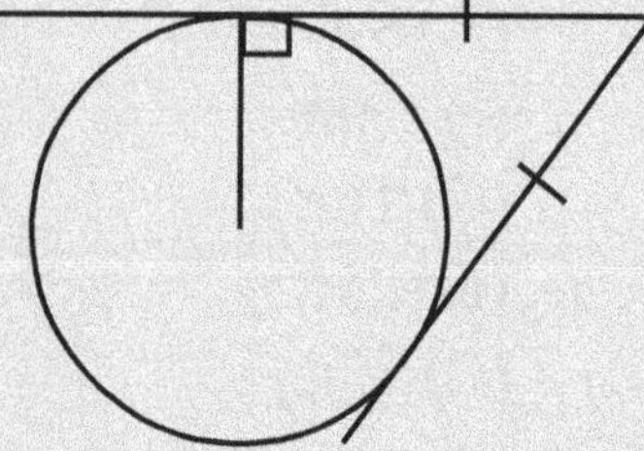

approach to sample questions

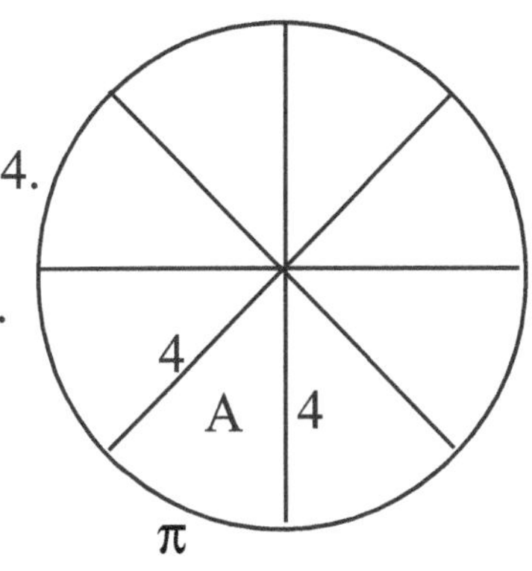

A1 Since the area of the circle is $\pi r^2 = 16\pi$, the radius is 4.
The circumference is $\pi D = 8\pi$.
Dividing 8π by 8, we see that the arc length of A is π.
Therefore, the perimeter of sector A is
$4 + 4 + \pi = 8 + \pi$.
The answer is (A).

A2 The area of sector AOB is $\frac{35}{360} \bullet 4\pi$
The area of sector COD is $\frac{x}{360} \bullet 2\pi$
Equating these two, we get $x = 70°$.
The answer is (C).

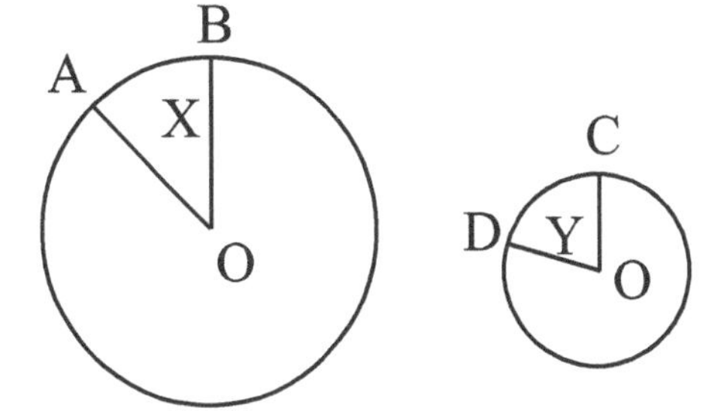

A3 The diagonal of the rectangle is the diameter of the circle.
It also forms a right triangle, and by Pythagorean rule, has a length of 10.
The radius, therefore, is 5 and the area is
$\pi r^2 = \pi(5)^2 = 25\pi$.
The answer is (D).

Practice Questions 7.9

1 What is the length of chord PQ if the radius of the circle is 10 and the length of arc PQ is 5π?

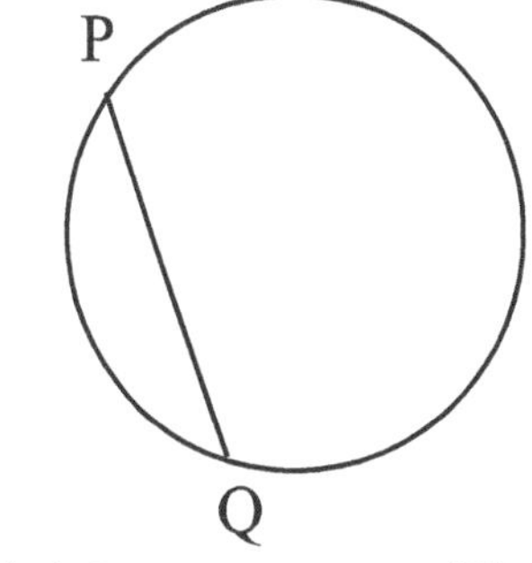

(A) 12.5 (B) 15 (C) $10\sqrt{2}$ (D) 16 (E) $\frac{12\pi}{5}$

2 Points A, B, C, and D are arranged on a circle in such a way that the four resulting chords AB, BC, CD, and DA are in descending order of lengths. Which of the following can be the degree measure of arc AB?

(A) 70
(B) 80
(C) 85
(D) 90
(E) 150

3 An equilateral triangle is inscribed in a circle of radius 5 and circumscribed about a circle of radius 3. Which of the following is the perimeter of the triangle?

(A) 12
(B) 15
(C) 18
(D) 24
(E) 30

4 In the figure below, if right triangle ABC is circumscribed about a circle and the lengths of AB and BC are 3 and 4, respectively, which of the following is the radius of the circle?

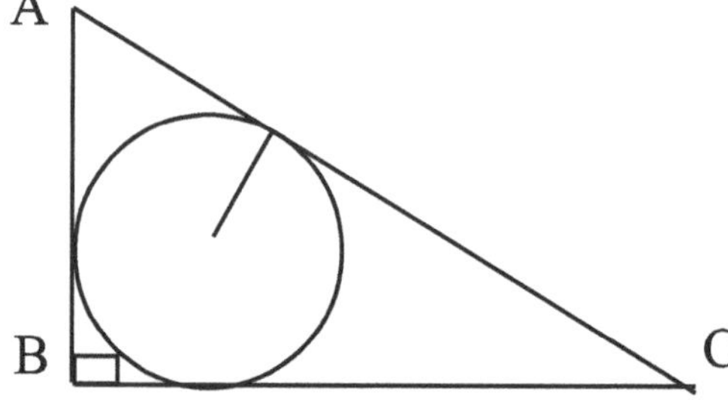

(A) 1/3
(B) 1/2
(C) 1
(D) $\sqrt{2}$
(E) 2

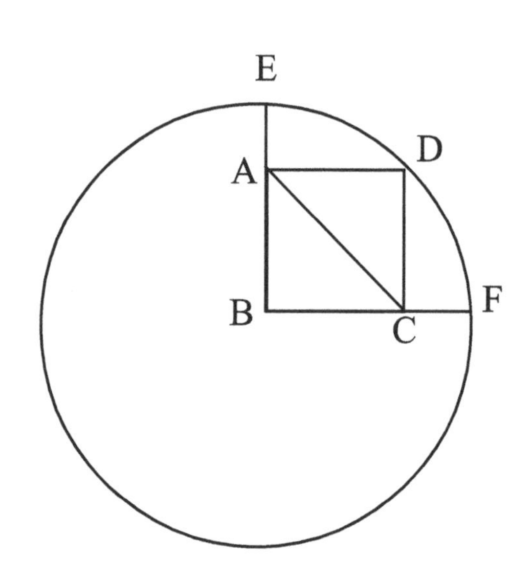

5 In the figure to the right, ABCD is a rectangle and EDF is a quarter circle with center B. If the length of arc EDF is 12π, what is the length of the diagonal AC?

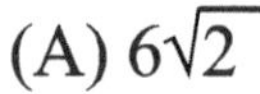

(A) $6\sqrt{2}$
(B) 12
(C) 24
(D) 36
(E) 48

7.10:: Surface Area / Volume / Similar Figures

try it yourself Try these three sample questions within 60 seconds.

Q1 Water fully contained in a cone-shaped container of radius 3 and height 7 is poured into an 11 x 3 x 5 rectangular container, shown in the figure below. What is the height of the water in the new container? (The volume of a cone is obtained by the fomula $V = (1/3)\pi r^2 h$, where r is the radius of the base, h is the height, and π is 22/7)

(A) 1
(B) 2
(C) 3
(D) 4
(E) 5

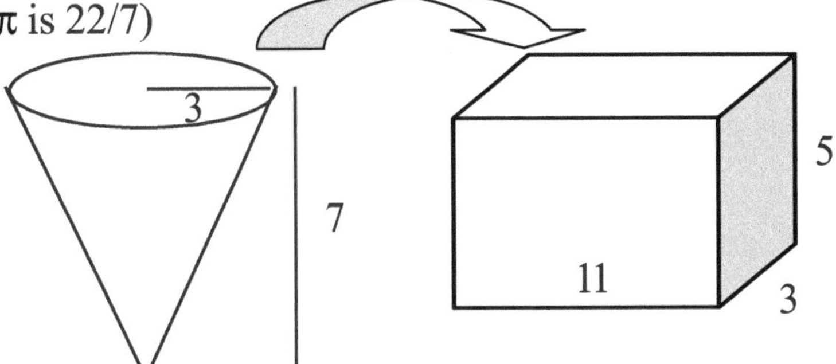

Q2 If the total surface area of a cube is 24 cubic inches, what is the volume of the cube, in cubic inches?

(A) 4
(B) 8
(C) 12
(D) 16
(E) 27

Q3 If the area of the large circle of radius r is 9k, and the radii of the three small circles are all r/3, what is the shaded area in terms of k?

(A) 4k
(B) 5k
(C) 6k
(D) 7k
(E) 8k

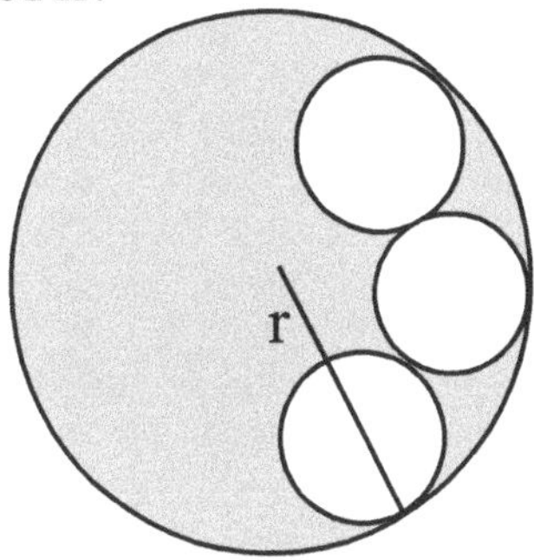

general rules

☑ **Surface area (S.A.) & Volume (V)**

Rectangular Cube & Cylinder
S.A. = 2 • (base area) + (lateral area) = $2(\pi r^2)(2\pi rh)$, or $6s^2$ for a cube
V = (base area) • height = lwh = $\pi r^2 h$

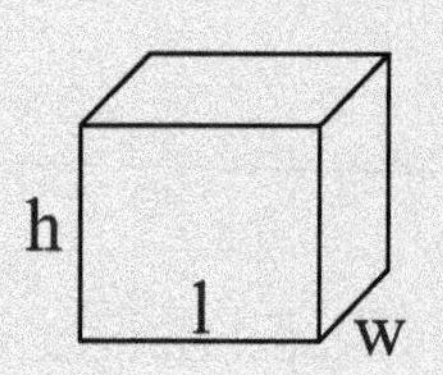

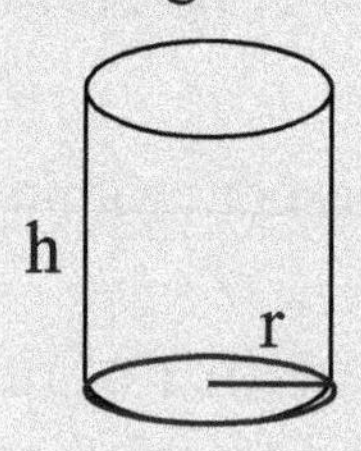

*Note that if you unravel a cylinder, it becomes a rectangle. The length of the rectangle is thus the circumference ($2\pi r$) of the base.

Cone
S.A. = (base area) + π • radius • length = $(\pi r^2) + \pi rl$
V = 1/3 • (base area) • height = $(1/3)(\pi r^2)h$

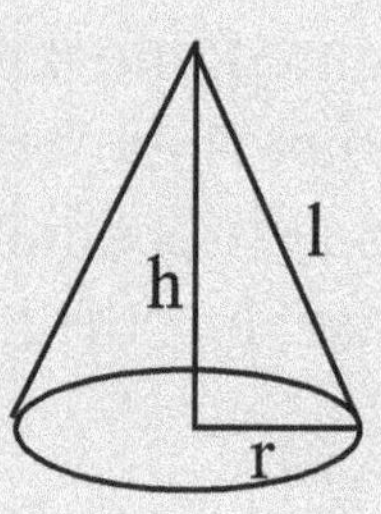

☑ **Diagonal of a cube**

The length of a **diagonal of a cube** is $\sqrt{3}$ times the length of an edge, s.
The length of a **diagonal of a square** is $\sqrt{2}$ times the length of an edge, s.

Remember the Pythagorean Theorem:
$a^2 + b^2 = c^2$

Similarly for the rectangular cube,
$l^2 + w^2 + h^2 = d^2$

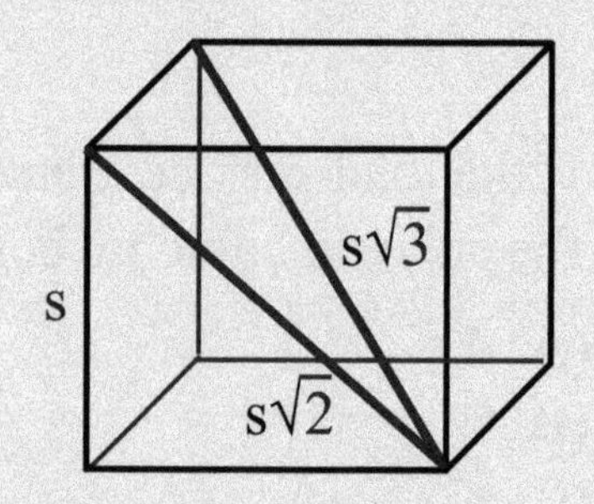

☑ **Very important relationships among length, area, and volume**
You do not need to know any fancy formulas!

Between any two similar figures, **if the length ratio between them is a:b**, then

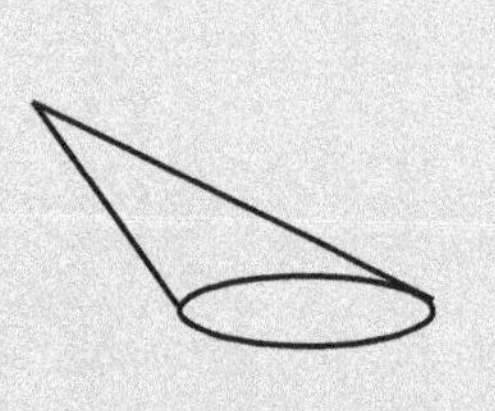

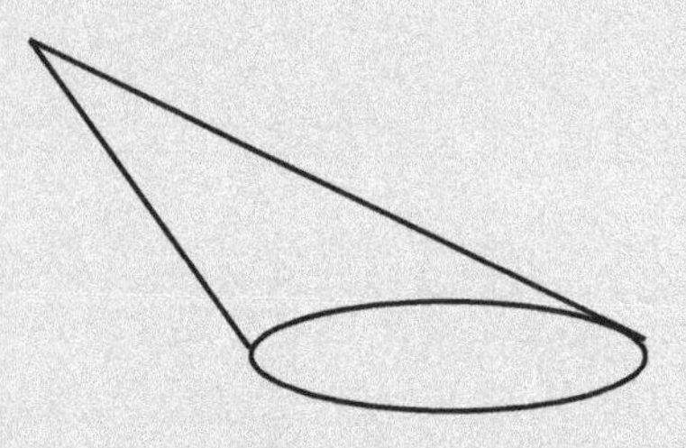

Area ratio is a^2:b^2 & **Volume ratio is a^3:b^3**

approach to sample questions

A1 The volume of the cone-shaped container is
$(1/3)(\pi r^2)h = (1)(22/3)(3)^2\,(7) = 66$.
Assuming the height of the water in the rectangular is h,
the volume of the water is $lwh = (11)(2)h = 22h$.
Since these volumes are equal, $66 = 22h$, and $h = 3$. The answer is (C).

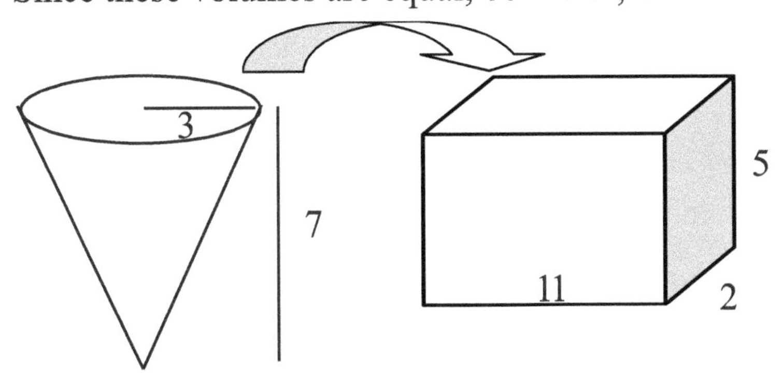

A2 Since the surface areas of the cube is 6 times the area of one face of the cube, $24 = 6x^2$, when x is the length of an edge.
Therefore, $x = 2$ and the volume is $2^3 = 8$. The answer is (B).

A3 The two types of circles of different sizes are similar.
Thus, the ratio of the radius is 3:1 and the area ratio is 9:1.
Therefore, if the area of the larger circle is 9k,
the sum of the areas of three smaller circles is 3k.
The shaded area will be 9k - 3k = 6k.
The answer is (C).

Practice Questions 7.10

1 What is the volume of a cube whose surface area is 0.24?

(A) 0.0012
(B) 0.008
(C) 0.012
(D) 0.04
(E) 0.2

2 Rectangular tank P measuring 4 x 11 x 7 is filled to the top with water. If the water is then poured into an empty cylindrical tank Q, which has a base radius of 3.5, what will be the height of the water in tank Q? (Use 22/7 for π).

(A) 4
(B) 6
(C) 8
(D) 10
(E) 12

3 In the rectangular box below with dimensions 1 by 2 by h, what is the square of the length of diagonal, d, in terms of h?

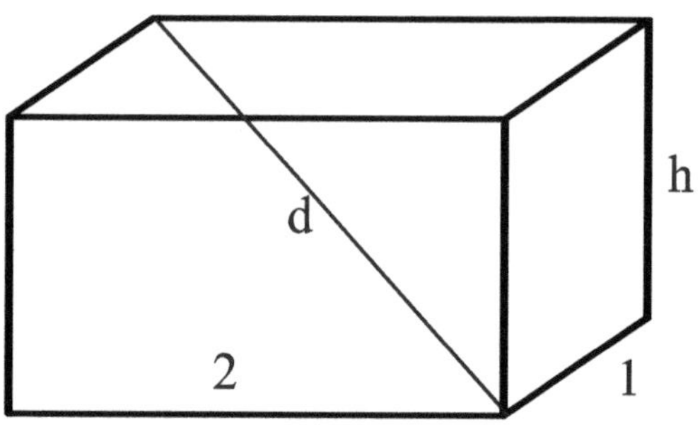

(A) $\sqrt{5} + h^2$
(B) $5 + h^2$
(C) $5 + h$
(D) $h\sqrt{5}$
(E) $\sqrt{5 + h}$

4 In the figure below, the circumference ratio between the two circles is 5 : 2. What is the ratio of the area of the shaded region to the area of the smaller circle?

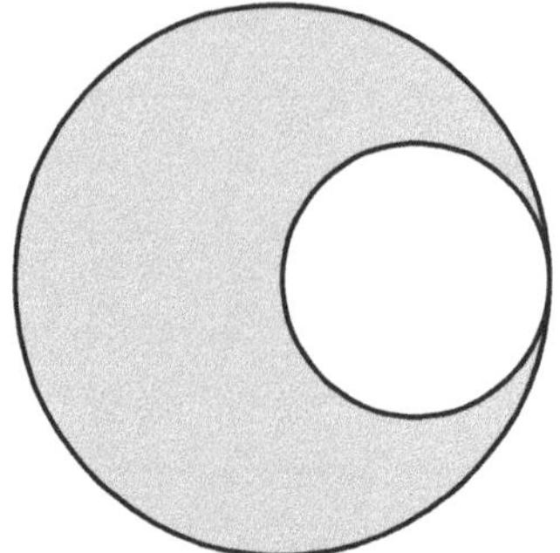

(A) 3 : 2
(B) 5 : 2
(C) 21 : 4
(D) 25 : 4
(E) 25 : 2

5 If the area ratio between the bases of two similar cones is 9 : 4, and the volume of the larger cone is 27k, what is the volume of the smaller cone?

(A) 4k
(B) 6k
(C) 8k
(D) 10k
(E) 16k

7.11:: Coordinate Geometry

try it yourself

Try this sample question within 30 seconds.

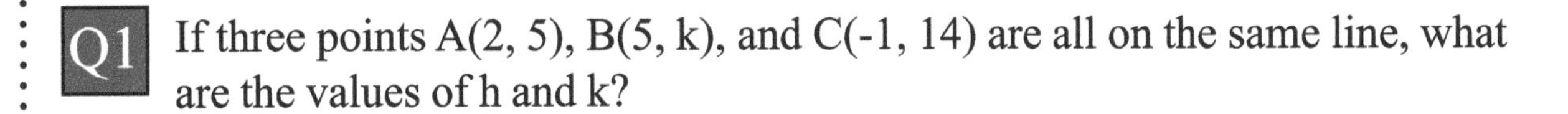

Q1 If three points A(2, 5), B(5, k), and C(-1, 14) are all on the same line, what are the values of h and k?

TIP: When working on the number line, pick actual approximate numbers to work with.

general rules

Between two points: P(a, b) and Q(c, d)

Q(c, d)
M
d - b
P(a, b)
c - a

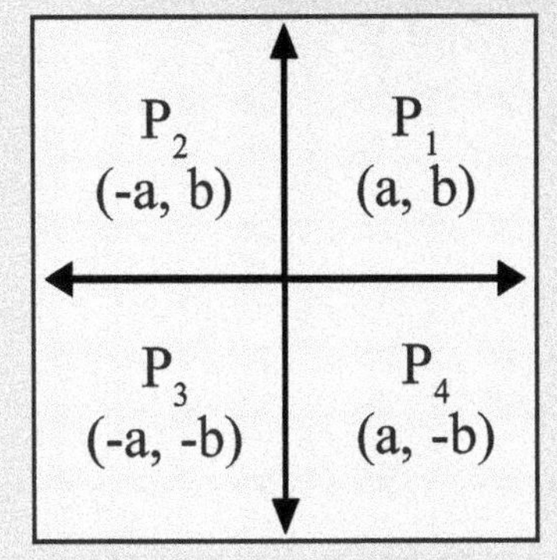

☑ **Midpoint**: $M\left(\frac{(a+c)}{2}, \frac{(b+d)}{2}\right)$

☑ **Slope**: $m = \frac{\text{rise}}{\text{run}} = \frac{(d-b)}{(c-a)}$

☑ **Circle** equation (center (0, 0) and radius r): $x^2 + y^2 = r^2$

☑ **Distance**: $PQ = d = \sqrt{(c-a)^2 + (d-b)^2}$

Most distance problems in the SAT can be solved by the **Pythagorean theorem,** using right triangles! $a^2 + b^2 = c^2$

☑ **Linear** equation (slope m & a point (c, d) are known)
Point-slope formula: $y - d = m(x - c)$

☑ **Perpendicular lines**: the product of the slopes is -1. $m_1 \bullet m_2 = -1$

approach to sample questions

A1 Formal approach:
Since all four points are on the same line, the slope of the line passing through any two points must be equal. Therefore,

$$\text{slope of AB} = \frac{y_2 - y_1}{x_2 - x_1} = \frac{k - 5}{5 - 2} = \frac{k - 5}{3} \qquad (1)$$

$$\text{slope of AC} = \frac{14 - 5}{-1 - 2} = \frac{9}{-3} = -3 \qquad (2)$$

$$\text{slope of CD} = \frac{2 - 14}{h + 1} = \frac{-12}{h + 1} \qquad (3)$$

Therefore, from (1) and (2), k - 5 = -9, and k = -4.
from (2) and (3), 4 = h + 1, and h = 3

Graphic approach:
Find a graph connecting two known points, A(2, 5) and C(-1, 14).
Then find two approximately corresponding x and y coordinates for x = 5 and y = 2.

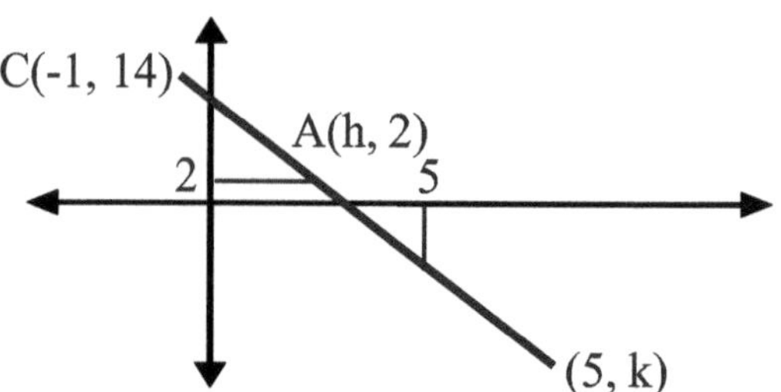

Practice Questions 7.11

1 In the figure of the circle with center O(0, 0) below, through which of the following points does it pass besides the point (-3, 0)?

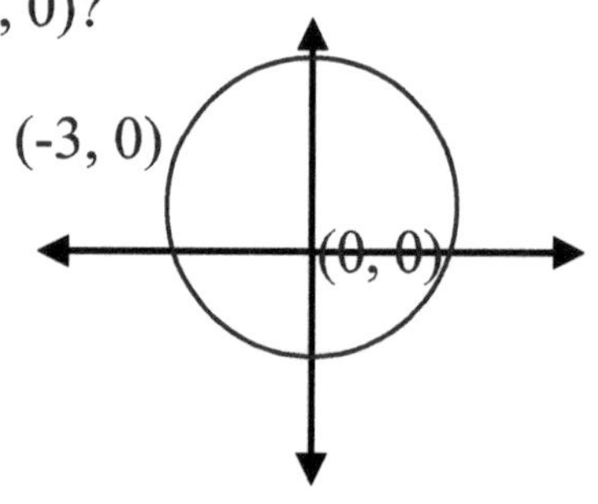

(A) $(1, \sqrt{7})$
(B) $(2, \sqrt{6})$
(C) $(-1, \sqrt{7})$
(D) (3,1)
(E) $(-2, -\sqrt{5})$

2

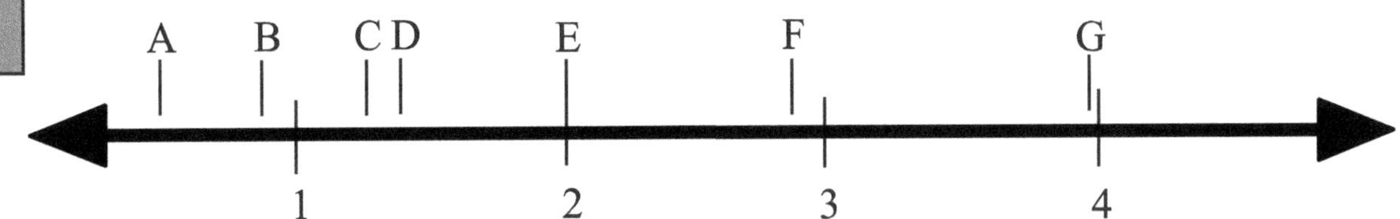

On the number line above, A/D is closest to which of the following ratios?

(A) B/E
(B) C/F
(C) D/G
(D) B/G
(E) C/G

3 ABCD is a parallelogram (not drawn to scale). What is coordinate B?

(A) (8, 10)
(B) (3, 8)
(C) (10, 10)
(D) (3, 10)
(E) (10, 3)

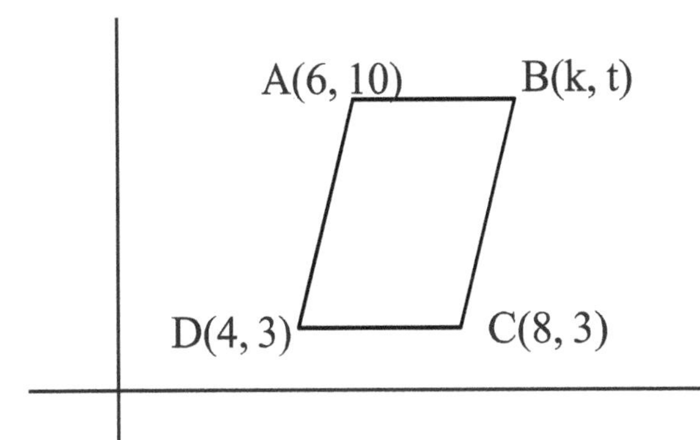

4 Which of the following points does not lie inside the triangle, shown below?

(A) (14, 7)
(B) (12, 6)
(C) (11, 6)
(D) (8, 3)
(E) (10, 5)

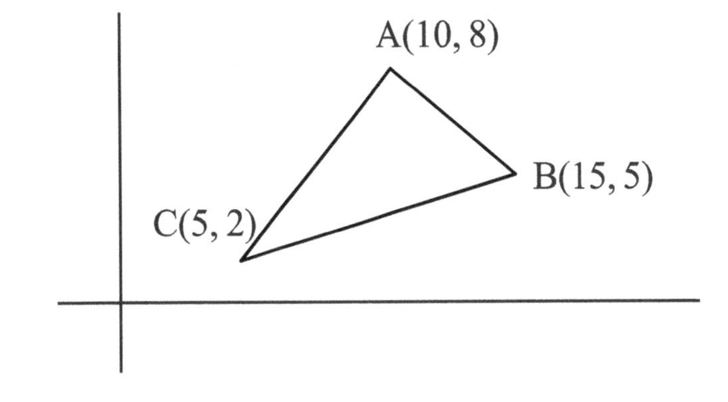

7.12:: Counting & Min/Max Search

try it yourself Try this sample question within 15 seconds.

Q1 How many triangles are formed by the five straight lines in the figure below?

general rules

☑ **Maximum area search**:
If the perimeter stays constant, the areas of polygons are maximum when there are as many sides as possible with lengths as close to each other as possible.

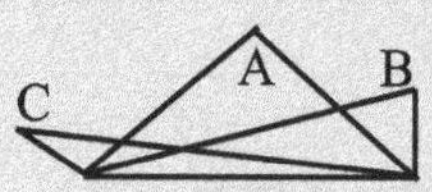

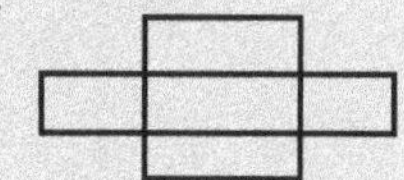
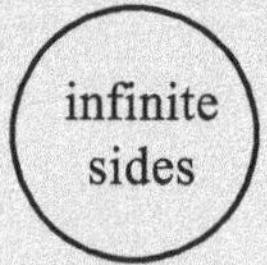

Area: A > B > C Square > Rectangle Circle

☑ **Maximum number of polygons search**:
Don't forget to count the polygons that are formed by multiple regions.

The figure has 5 regions, but 9 triangles.

☑ **Maximum number of intersections search:**
Only through trial and error drawing.

☑ **Maximum number of paths search**:
This is related to probability, but you can trace all paths out without using a formula.

☑ **Maximum number of non-overlapping regions search:**
Only through trial and error drawing.

☑ **Maximum distance search**:
Depends on the order and direction.

approach to sample questions

A1 First, count the single-region triangles : 6
Then, the multiple-region triangles : 5
There are 11 triangles in the figure.

Practice Questions 7.12

1 In the rectangle below, if x and y are integers, $x + y = 15$ and $x > y$, what would the value of x have to be in order to obtain the maximum possible area?

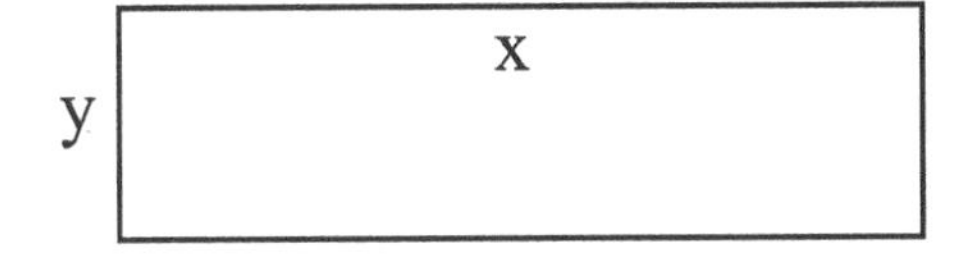

(A) 8
(B) 9
(C) 10
(D) 13
(E) 14

2 The figure below is made up of 6 parallel lines. What is the total number of different parallelograms formed by these lines?

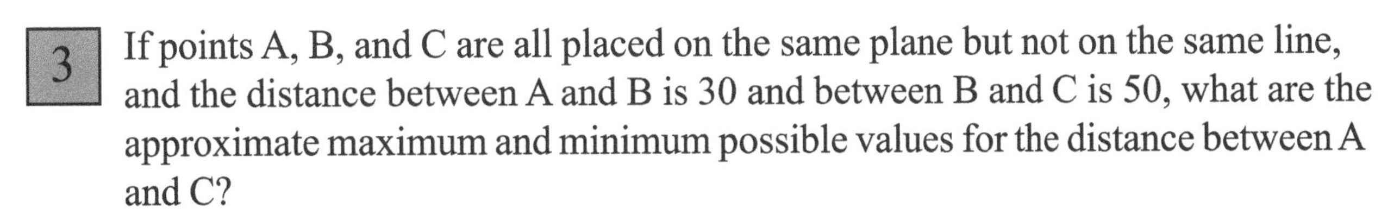

(A) 4
(B) 5
(C) 8
(D) 9
(E) 10

3 If points A, B, and C are all placed on the same plane but not on the same line, and the distance between A and B is 30 and between B and C is 50, what are the approximate maximum and minimum possible values for the distance between A and C?

(A) 69 and 11
(B) 75 and 15
(C) 79 and 21
(D) 89 and 31
(E) 85 and 25

4 A triangle and a square both have the same perimeters of 24 inches. Based on this information, which of the following statements is true?

(A) The maximum possible area of the triangle is greater than the maximum possible area of the square.
(B) The maximum possible area of the square is greater than the maximum possible area of the triangle.
(C) The maximum possible area of the triangle is equal to the maximum possible area of the square.
(D) The maximum possible area cannot be determined based on this information.

5 Four lines divide a square region into a maximum of how many non-overlapping triangles?

(A) 5
(B) 6
(C) 7
(D) 8
(E) 9

6 In the grid of unit squares below, there are multiple possible paths that can be traced along the lines from A to T via R. If one wants to take the shortest possible route from A to T, how many possible paths can he follow?

(A) 20
(B) 30
(C) 40
(D) 50
(E) 60

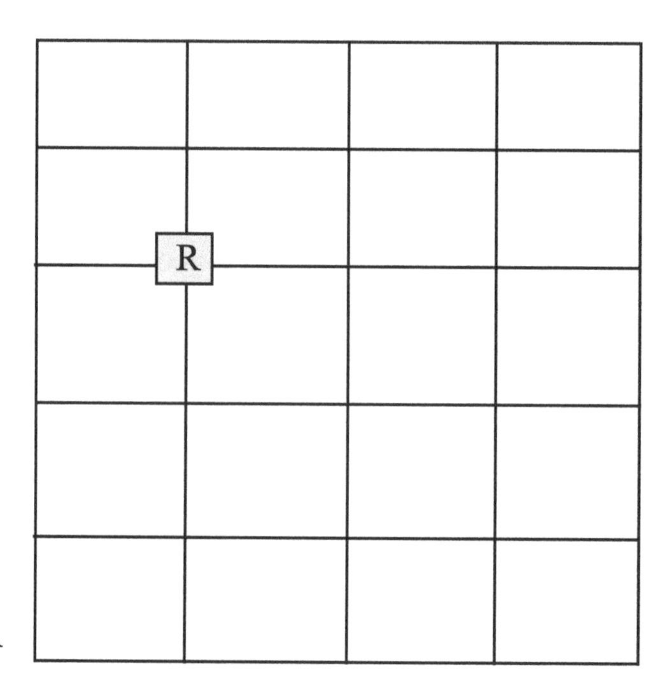

7.13:: Algebra in Geometry

try it yourself

Try these three sample questions within 60 seconds.

Q1 If the area of the shaded region in the figure below is 27π, what is the radius r of the smaller circle?

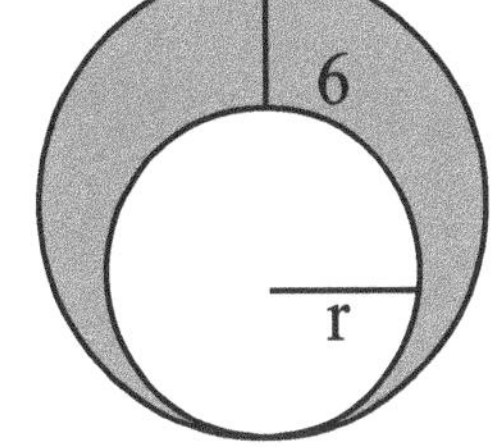

(A) 1
(B) 1.5
(C) 2
(D) 2.5
(E) 3

Q2 If a circle has the same area as a rectangle with width $1/\pi$ and length $4/\pi$, what is the radius of the circle?

(A) $\frac{\sqrt{\pi}}{2\pi}$ (B) $\frac{\sqrt{\pi}}{\pi^2}$ (C) $\frac{2\sqrt{\pi}}{\pi^2}$ (D) $\frac{\sqrt{\pi}}{\pi}$ (E) $\frac{4}{\pi}$

Q3 SO and OT are perpendicular and $x > 40°$ (figure not drawn to scale). Which of the following statements is true?

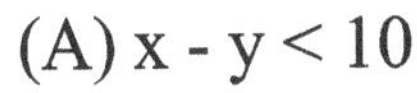

(A) $x - y < 10$
(B) $x - y = 10$
(C) $x - y > 10$
(D) $x - y = 90$
(E) $x - y > 90$

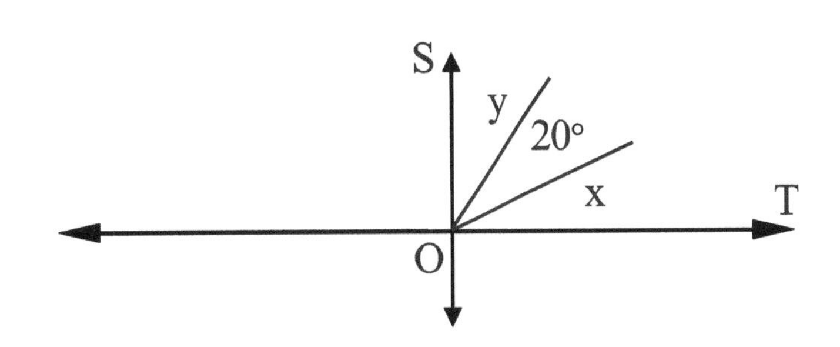

Treat **variables as numbers** when applied to geometric formulas or engaged in any other algebraic operation.

general rules

approach to sample questions

A1 Polynomial product:

Since the diameter of the big circle is 2r + 6, the radius is r + 3.

Area of the shaded region = (Area of big circle) - (Area of small circle)

$27\pi = \pi(3 + r)^2 - \pi r^2$

$27\pi = \pi(9 + 6r + r^2) - \pi r^2$

$27\pi = 9\pi + 6\pi r + \pi r^2 - \pi r^2$

$6\pi r = 18\pi$

r = 3. The answer is (E).

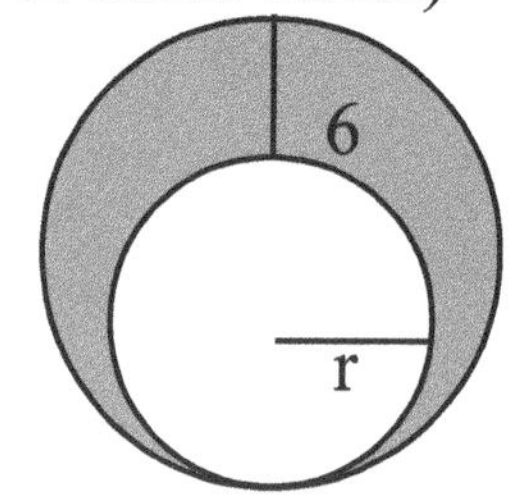

A2 Radical operations:

Area of a circle = πr^2

Rectangle area = $4/\pi^2$

$\pi r^2 = 4/\pi^2$

$r^2 = 4/\pi^3$

$r = \dfrac{2}{\pi\sqrt{\pi}} = \dfrac{2\sqrt{\pi}}{\pi^2}$. The answer is (C).

A3 Inequality operations:

Since y + 20 + x = 90

x + y = 70

x = 70 - y.

Therefore, 70 - y > 40; 30 > y.

$$\begin{array}{rl} \text{Since } & 40 < x < 70 \\ - & 0 < y < 30 \\ \hline & 10 < x - y < 70 \end{array}$$

(see cross subtraction in inequality operations, Section 3.4)

Or, you can use border numbers such as 40 for x and 30 for y.

The answer is (B).

Practice Questions 7.13

1 What is the area of a circle with circumference $2/\pi$?

(A) $1/\pi^2$ (B) $1/(2\pi)$ (C) $1/\pi$ (D) $2/\pi$ (E) 2π

2 What is the length of one edge of a cube whose volume is 2h cubic inches and surface area is 3h square inches?

(A) 1
(B) 2
(C) 4
(D) 6
(E) 8

3 In the triangle below, if $40 < y < 60$, which of the following must be true? (figure not drawn to scale)

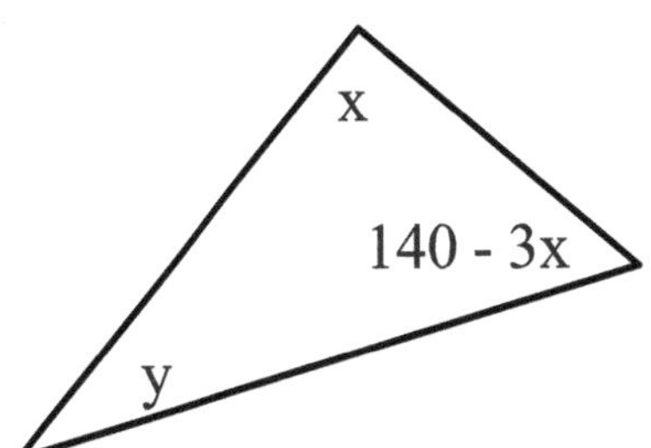

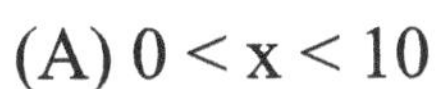

(A) $0 < x < 10$
(B) $10 < x < 20$
(C) $20 < x < 30$
(D) $0 < x < 20$
(E) $10 < x < 30$

4 The figure to the right shows the areas for each of the two regions of a circle. What percent of the circle is xy?

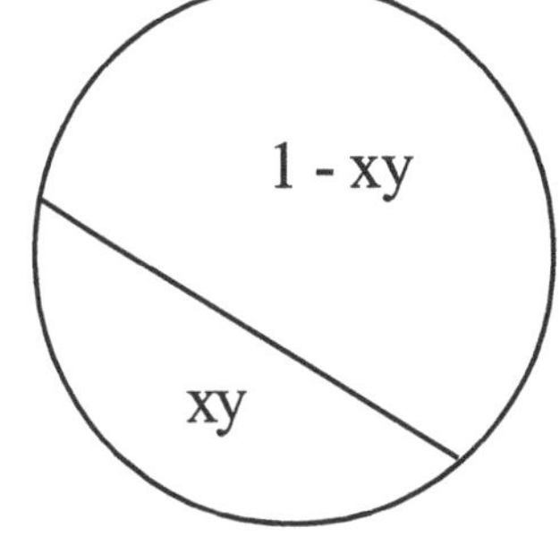

(A) xy%
(B) xy/100%
(C) 100xy%
(D) 100/xy%
(E) 100x/y%

5 Square ABCD with side h, below, is cut into two non-overlapping regions by line EF, which is parallel to diagonal AC in the figure. If the shaded area is 1/3 of the non-shaded area, what is the length of EF in terms of h?

(A) $h(\sqrt{3})/4$
(B) $h(\sqrt{2})/3$
(C) $h/2$
(D) $h(\sqrt{2})/2$
(E) $h(\sqrt{3})/2$

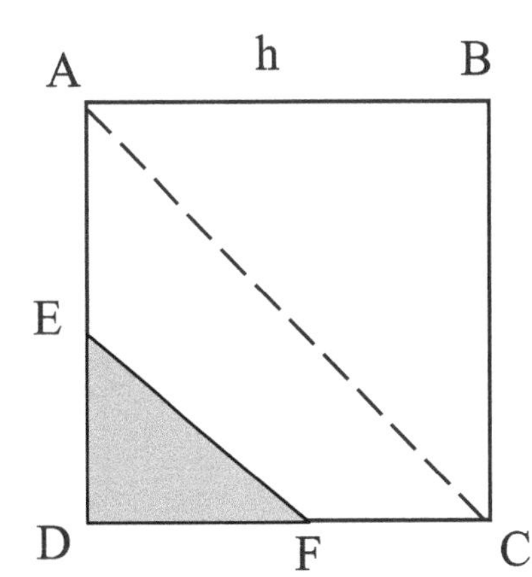

7.14:: Test of Imagination / Drawing

try it yourself Try these two sample questions within 30 seconds.

Q1 If line segment XY, shown below, is 4, what is the distance traveled by the midpoint M when segment XY is rotated 270 degrees about point X?

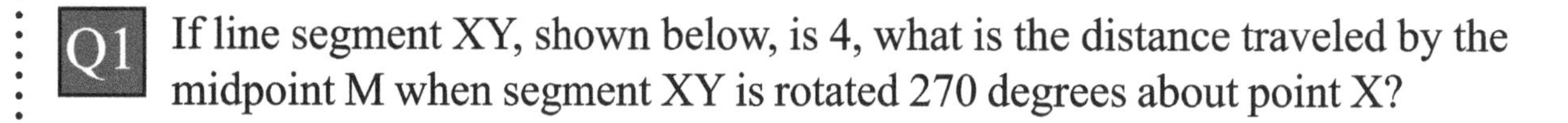

Q2 Circle O with radius 4 overlaps with Circle O', with radius 3. Points A and B mark the two points of intersection between the circles. If $OA \perp O'B$, what is the length of OO'?

general rules

The general rule is to **draw** all the possible cases on paper instead of keeping them as mental pictures!

☑ Drawing can solve most of the problem for us.

☑ **Label** the picture according to the specifications provided.

☑ **Do not assume anything** about the picture. Beware of the tendency to read things as we wish.

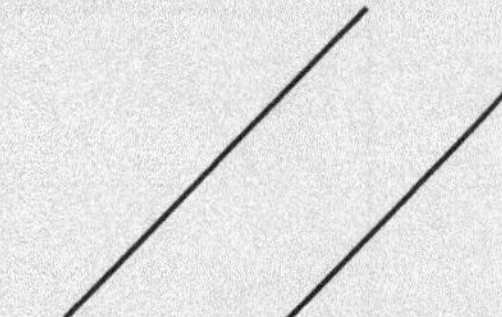

Are these lines parallel?

Is this figure a square?

No, **unless the question states so.**

approach to sample questions

A1 Rotating XY 270 degrees about X creates a 3/4 circle, with X as the center, XY as the radius, and M as the midpoint of the radius. XM, then, has the length of 2. Therefore, the length of the trace is 3/4 of the circumference of the entire circle of radius 2. Since the circumference is πD, the trace is $3/4 \bullet 4\pi = 3\pi$.

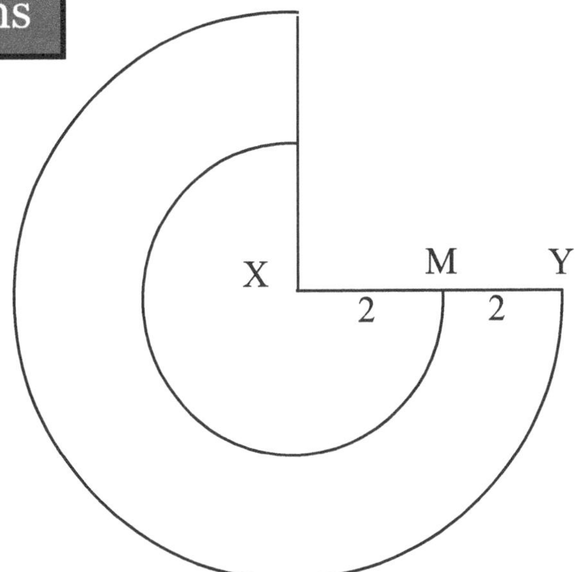

A2 According to the specifications, the triangle AOO' is a right triangle. Therefore, using the Pythagorean theorem, OO' = 5.

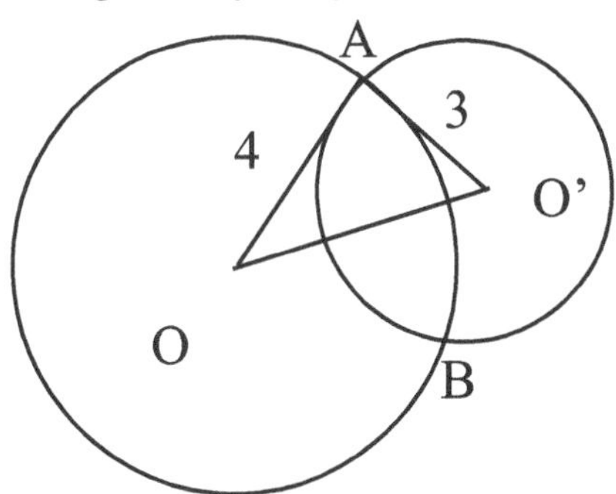

Practice Questions 7.14

1 In a circle of radius 4, what is the area of the larger segment left after cutting out a smaller segment along a chord that is $4\sqrt{2}$ long?

(A) $8\pi + 4\sqrt{2}$
(B) $8\pi + 8$
(C) $12\pi + 8$
(D) $12\pi + 16$
(E) $16\pi + 4\sqrt{2}$

2 What is the area of the largest triangle formed by connecting 3 vertices of a cube with an edge of 4 inches?

(A) $4\sqrt{2}$
(B) $8\sqrt{2}$
(C) $8\sqrt{3}$
(D) $16\sqrt{2}$
(E) $16\pi + 16$

3 Quadrilateral VWXY has two equal adjacent acute angles at V and W. Which of the following statements must be true?

(A) $XV < YW$
(B) $\angle Y$ is acute
(C) Either $\angle X$ or $\angle Y$ is obtuse
(D) $YV = XW$
(E) $YX < VW$

4 Points V, W, X, and Y lie on a line in that order. If VX/VY = 2/3 and VY/VW = 2/1, what is the value of WY/VX?

CHAPTER 7 TEST

1 Which of the following is not true about angle x in the figure below?

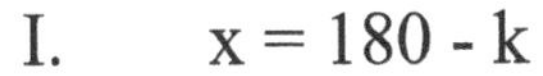

I. $x = 180 - k$

II. $x = 180 - (h + m)$

III. $x = h + m$

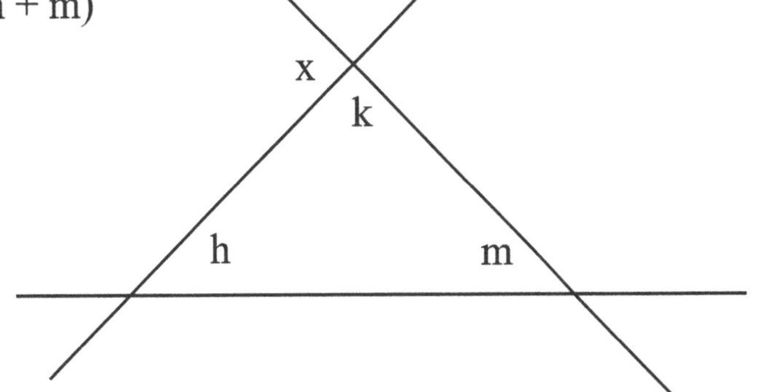

(A) I only
(B) II only
(C) III only
(D) I and III
(IV) I, II, and III

2 If lines L_1 and L_2 are parallel but lines M_1 and M_2 are not, which of the following about the angles is true?

I. $a + b = 180$

II. $a + d = 180$

III. $b = c$

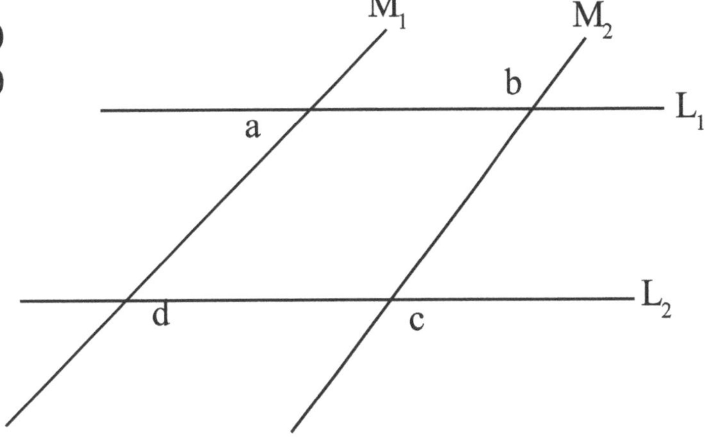

(A) I only
(B) II only
(C) III only
(D) I and III
(E) II and III

3 Which of the following statements about the figure below is not true if AD and BC are perpendicular?

(A) ΔABC is a right triangle
(B) ΔADC is an isosceles triangle
(C) $m\angle z = 45°$
(D) ΔADC is a right triangle
(E) $m\angle x + m\angle y = 2m\angle z$

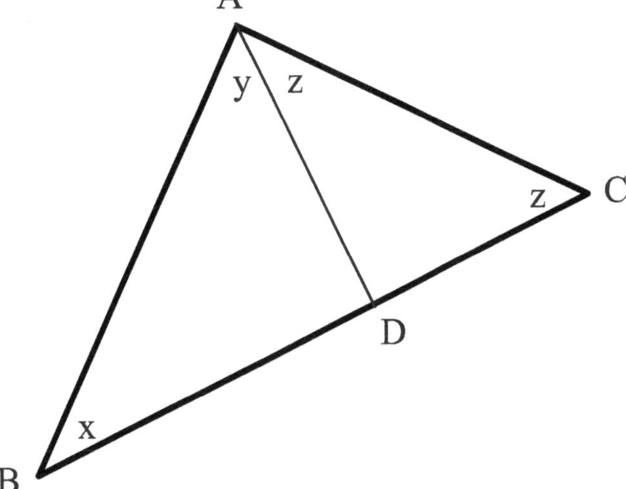

In right triangle ABC, AC and BD are perpendicular, $m\angle C = 30°$, and BC is x. What is the length of AD, in terms of x?

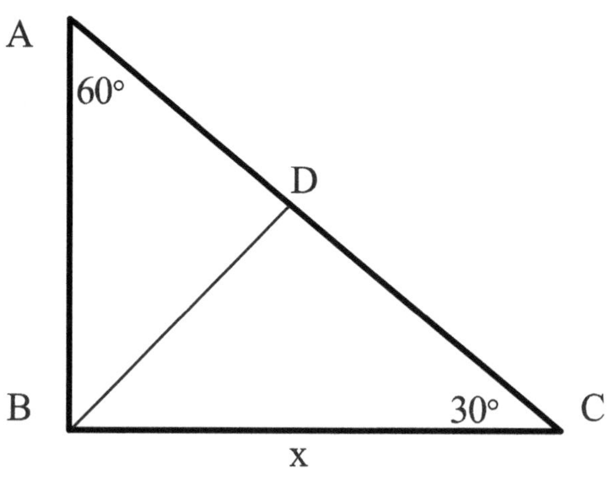

(A) $x\sqrt{3}$
(B) $x\sqrt{3}/2$
(C) $\sqrt{2}/(3x)$
(D) $x\sqrt{3}/6$
(E) $2x/\sqrt{3}$

5

If the sides of a triangle are 3, 7, and x, how many integer values are possible for the value of x?

(A) 3
(B) 5
(C) 7
(D) 9
(E) 15

6

How many revolutions does a bike wheel of radius k meters make in traveling a distance of 500 meters?

(A) $500/k\pi$
(B) $500k/\pi$
(C) $250\pi/k$
(D) $250\pi/k$
(E) $250/k\pi$

7

If an isosceles right triangle has the area 32, what is the perimeter of the triangle?

(A) $8 + 4\sqrt{2}$
(B) $16 + \sqrt{2}$
(C) $8 + 8\sqrt{2}$
(D) $16\sqrt{2}$
(E) $16 + 8\sqrt{2}$

8 If the area of the square inscribed in the quarter circle is x^2, what is the perimeter of regions C + A?

(A) $x\sqrt{2}(1 + \pi/2)$
(B) $x\sqrt{2}(2 + \pi/4)$
(C) $x\sqrt{2}(2 + \pi/2)$
(D) $x\sqrt{2}(1 + \pi)$
(E) $x\sqrt{2}(2 + 2\pi)$

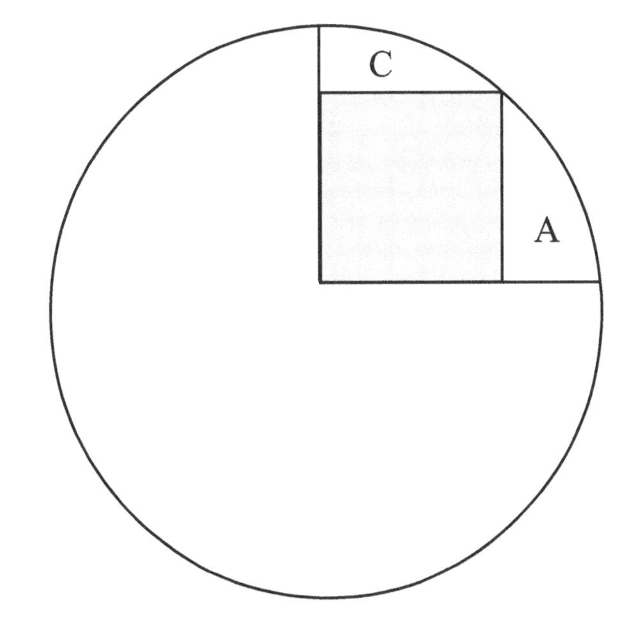

9 The two concentric circles below have radii $\sqrt{2}$ and $\sqrt{3}$ inches, respectively. If arc AB is 90°, what is the length of chord CD?

(A) $\sqrt{2}$
(B) $\sqrt{3}$
(C) 2
(D) $2\sqrt{2}$
(E) $2\sqrt{3}$

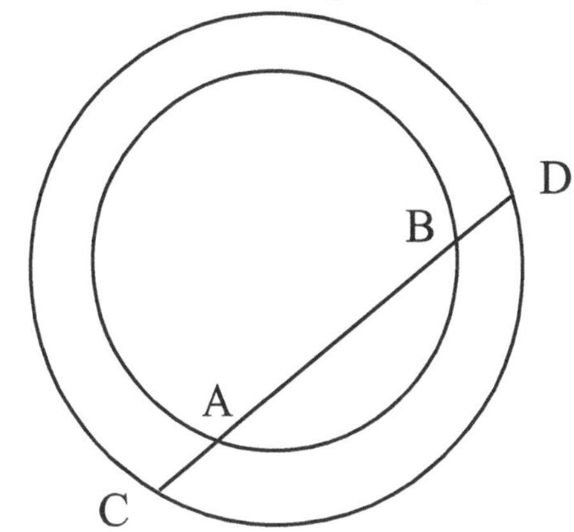

10 If the ratio of the surface areas of two spheres is 4 : 9, and the volume of the smaller sphere is 16, what is the volume of the larger sphere?

(A) 24
(B) 36
(C) 48
(D) 54
(E) 72

12 Four lines can divide an equilateral triangle into a maximum of how many nonoverlapping regions?

(A) 7
(B) 8
(C) 10
(D) 11
(E) 12

13 If the length of the diagonal of a cube is 3, what is the surface area of the cube?

(A) $\sqrt{3}$
(B) $3\sqrt{3}$
(C) $12\sqrt{3}$
(D) 12
(E) 18

14 Which of the following is true for isosceles triangle ABC, where AB = AC, altitude BE ⊥ AC, altitude AD ⊥ BC, and A = 58°?

(A) AB < BC
(B) BD > AE
(C) BE < AD
(D) AC = BC
(E) ∠C < ∠A

ANSWER KEY

For Practice Questions & Chapter Tests

Practice Questions 1.1

1. (5); $7(x + y) - 4(x + y) = 15$
$3(x + y) = 15$
$x + y = 5.$

2. (28); Use binomial factoring:
$x^2 - y^2 = (x + y)(x - y) = (7)(4) = 28.$

3. (-2); $x^2 - xy = x(x - y) = 24$
$6x = 24; x = 4$
Then $x - y = 6$
$4 - y = 6; y = -2$

4. (99); $80(11)^2 + (11)^2 = 81(11)^2 = (9)^2(11)^2$
$\sqrt{(9)^2(11)^2} = 99.$

5. (7^{16}); Use monomial factoring:
$\frac{7^{17} - 7^{16}}{6} = \frac{7^{16}(7 - 1)}{6} = \frac{7^{16}(6)}{6} = 7^{16}.$

6. (D); $\frac{10! - 9!}{9} = \frac{9!(10 - 1)}{9} = \frac{9!(9)}{9} = 9!.$
Therefore, the answer is (D).

Practice Questions 1.2

1. (4); Remember, we are not looking for x or y separately, but x + y as a whole.
Adding the two equations:
$3x + 4y = 7$
$2x + y = 13$
$5x + 5y = 20$
Therefore, $x + y = 4$.

2. (5); Subtracting one equation from the other:
$4x - 3y = 10$
$-3x - 4y = 5$
$x + y = 5.$

3. (4); Adding the three equations:
$a + b = 7$
$b + c = 12$
$c + a = 5$
$2a + 2b + 2c = 24$
Therefore, $a + b + c = 12$.
The arithmetic mean of a, b, and c is
$\frac{a + b + c}{3} = \frac{12}{3} = 4.$

4. (5); Adding the three equations:
$x + y + z = 18$
$y + z + t = 15$
$x + t = 7$
$2x + 2y + 2z + 2t = 40$
Therefore, $x + y + z + t = 20$.
The average of x, y, z, and t is
$\frac{x + y + z + t}{4} = \frac{20}{4} = 5.$

5. (1); In order to obtain $-3a + 2b$, subtract the first equation from the second:
$a + 3b = 23$
$-4a - b = -22$
$-3a + 2b = 1.$

Practice Questions 1.3

1. (2); Dividing the first equation by the second:
$\frac{mn(x + y) = 10}{n(x + y) = 5}$
Therefore, $m = 2$.

2. (8/9); Flip a/c to become c/a so that
$\frac{a}{b} \bullet \frac{c}{a} = \frac{c}{b}.$
Therefore, $\frac{4}{3} \bullet \frac{2}{3} = \frac{8}{9}.$

3. (5); Divide the first equation by the second:
$\frac{xy^2z^3}{yz^2} = \frac{100}{20}$
Therefore, $xyz = 5$.

4. (6/5); Divide the first equation by the second:
$\frac{3(a - b) = 2k^2}{5(a - b) = 4k}$
Then $\frac{3}{5} = \frac{2k^2}{4}$ and $5k = 6$.
Use cross multiplication and rectangular cancelation (page 64) to obtain
$\frac{3}{5} = \frac{k}{2}$ and $5k = 6$
Therefore $k = 6/5$.

5. (1,000); Multiply the three equations together:
$xy = 5$
$yz = 4$
$zx = 5$
$x^2y^2z^2 = 100 = (xyz)^2.$
If $z > 0$, x and y both have to be positive in order for their product to be positive, which means $xyz = 10$.
Therefore, $x^3y^3z^3 = 10^3 = 1{,}000.$

Practice Questions 1.4

1. (6); Break up the equation:
$\frac{x^2 + y^2}{xy} = \frac{x^2}{xy} + \frac{y^2}{xy} = \frac{x}{y} + \frac{y}{x} = 10$
Therefore, $\frac{x}{y} + 4 = 10$, and $\frac{x}{y} = 6$.

2. (0); Remove the parentheses:
$3/4 + 4/5 - 4/5 - 5/6 + 5/6 - 3/4 = 0.$

3. (8); Break up the equation:
$\frac{ab + b^2}{b} = \frac{ab}{b} + \frac{b^2}{b} = 15$
Therefore, $a + b = 15$.
Since $b = 7$, $a + 7 = 15$ and $a = 8$.

4. (1/5); Combine the first equation:

$$\frac{a}{b} + \frac{b}{a} = \frac{a^2}{ab} + \frac{b^2}{ab} = 15$$

$$\frac{a^2 + b^2}{ab} = 15$$

Since $a^2 + b^2 = 3$, $3/ab = 15$ and $ab = 1/5$.

5. (3/40); Break up the expression:

$$\frac{b - a}{ab} = \frac{b}{ab} - \frac{a}{ab} = \frac{1}{a} - \frac{1}{b}$$

The question is asking for the maximum value, which means the first number should be as large as possible (small denominator) while the second should be as small as possible (large denominator).

If a = 5 and b = 8, $\frac{8 - 5}{(8)(5)} = \frac{3}{40}$.

Practice Questions 1.5

1. (3a + b); Divide both the numerator and denominator by 2.

2. (-1); $x - y = -(y - x)$, so $\frac{-(y - x)}{y - x} = -1$.

3. (-24); $14y - 10x = 2(7y - 5x)$
$= -2(5x - 7y)$
Substitute: $= -2(12)$
$= -24$.

4. (-6); Transform $6b - 3a$ to $-3(a - 2b)$

$$\frac{-3(a - 2b)}{t} = \frac{-3(2t)}{t} = -6.$$

5. (E); When the numerator and denominator are multiplied by the same number (x/x = 1), the value of the fraction is the same.

$$\frac{2}{\frac{2+2}{x}} = \frac{1 \quad (2)}{\frac{1+1}{x} \; (2)}$$

Therefore, the answer is (E).

Practice Questions 1.6

1. (2t); If t = 3 + 6 + 9 + ... + 96 + 99, and
s = 6 + 12 + 18 + ... + 192 + 198,
you'll notice that every term is doubled.
Therefore, s in terms of t is 2t.

2. (2k + 250,000); If 1 + 2 + ... + 499 + 500 = k,
then 501 + 502 + ... + 999 + 1,000
= (500 + 1) + (500 + 2) + ... + (500 + 500)
= (500)(500) + 1 + 2 + ... + 499 + 500
= 250,000 + k
Therefore, 1 + 2 + ... + 499 + 500 + 501 + ... + 999 + 1,000 = k + k + 250,000 = 2k + 250,000.

3. (500); The sum of the odd integers subtracted by the sum of the even integers is written below:

E:	2 + 4 + 6 + ... + 998 + 1000
-O:	1 + 3 + 5 + ... + 997 + 999
	1 + 1 + 1 + ... + 1 + 1 = 500.

4. (3,775); $\frac{100(101)}{2} - \frac{50(51)}{2} = \frac{10,100 - 2,550}{2}$
$= \frac{7,550}{2} = 3,775$.

5. (62,500); Write the first and last parts of both sequences:

u:	251 + 252 + ... + 499 + 500
-t:	1 + 2 + ... + 249 + 250
	250 + 250 + ... + 250 + 250

Therefore, u - t = (250)(250) = 62,500.

Chapter 1 Test

1. (B); $x^2 - y^2 = (x + y)(x - y) = (4)(2) = 8$.

2. (C); Adding both equations:
2x + 3y = 10
3x + 27 = 10
5x + 5y = 20; Therefore x + y = 4.

3. (A); 2 + 4 + 6 + ... + 100 is
(1 + 1) + (3 + 1) + (5 + 1) + ... + (99 + 1)
= 1 + 3 + 5 + ... + 99 + 50
= 2,500 + 50 = 2,550.

4. (A); $x^2 + xy = x(x + y) = 84$
12x = 84; x = 7
Then, x + y = 12, so 7 + y = 12 and y = 5.

5. (E); Subtracting the two equations:
3x + y = 15
-x - 2y = 16
2x - y = -1.

6. (D); t = 31 + 32 + 33 + ... + 59 + 60
-s = 1 + 2 + 3 + ... + 29 + 30
t - s = 30 + 30 + 30 + ... + 30 + 30
= (30)(30) = 900.

7. (C); a - b = 5 = -(b - a).

8. (B); Since the questions asks about b, subtract the second equation from the first:
$a^2 + b^2 = 41$
$-a^2 + b^2 = -9$
$2b^2 = 32$
Then, $b^2 = 16$ and $b = \pm 4$.
Since b is a positive integer, b = 4.

9. (E); Multiply the three equations together:

$$\begin{aligned} zx &= 150 \\ yz &= 20 \\ xy &= 30 \\ \hline x^2y^2z^2 &= 90{,}000, \text{ so } xyz = 300. \end{aligned}$$

10. (D); Transform $3a - 4b = 4b - 3a = 6$.
Then, $8b - 6a = 2(4b - 3a) = 2(6) = 12$.

11. (1,600 + x); If $x = 1 + 2 + ... + 39 + 40$,
Then $41 + 42 + ... + 79 + 80$
$= (40 + 1) + (40 + 2) + ... (40+ 39) + (40 + 40)$
$= (40)(40) + 1 + 2 + ... + 39 + 40$
$= 1{,}600 + x$.

12. (E); $\frac{x}{z} = \frac{(x)}{(y)} \bullet \frac{(y)}{(z)} = \frac{k}{3} \bullet \frac{(k)}{3} = k^2$.

Practice Questions 2.1

1. (E); The following are examples that validate each statement:
I: $1.5 - 0.5 = 1$, which is an integer
II: $2.5/0.5 = 5$, which is an integer
III: $\sqrt{4.5 - 0.5} = 2$, which is an integer.

2. (A); If s is not an integer, dividing s itself by an integer can never result in an integer.

3. (C); There are two possible cases for $x(m - n)$ to be negative:
II: If x is positive, $m - n$ must be negative;
$m - n < 0$ and $n > m$
III: If x is negative, $m - n$ must be positive;
$m - n > 0$ and $m > n$
Therefore, both II and III are correct.

4. (E); If a and b are negative integers and $a < b$, an example of this would be $a = -4$ and $b = -2$, in which case a/b would equal $-4/-2 = 2$.

5. (D); The only possible values for s and t are 1 and -1, respectively, in which case
$st = (1)(-1) = -1$.

Practice Questions 2.2

1. (D); An even number divided by 2 can be both even or odd, depending on the value of G..
However, (even) - (odd) = odd
and $(\text{odd})^2 = (\text{odd})(\text{odd}) = \text{even}$.

2. (B); In order for the number to be divisible by 4 and 5, the last two-digit number k0 has to be divisible by 20. At the same time, in order for the number to divisible by 3 and 9, the sum of all its digits has to be divisible by 3 and 9.
$6 + 7 + 9 + 3 + k = 27$
Therefore, $k = 2$.

3. (C); Since we do not know whether G is even or odd, we must consider both cases:
I: If G is even: $G + 2$ is even, $3G + 1$ is odd,
$4G - 1$ is odd, and $2(2G - 1)$ is even.
II: If G is odd: $G + 2$ is odd, $3G + 1$ is even,
$4G - 1$ is even, and $2(2G - 1)$ is odd.
In both cases, there are 2 odd and 2 even integers.

4. (E); All statements can be true.
I: The multiples of 11 can end in 1 (11), 3 (33), and 9 (99).
II: Numbers that end in 1, 3, and 9 can be prime numbers as well (11, 13, 19).
III: Powers of 7 can end in 1, 3, and 9; for example, $7^2 = 49$, $7^3 = ..3$, and $7^4 = ..1$.

5. (D); In order for $xy(y - z)$ to be an odd number, x, y, and $(y - z)$ all have to be odd numbers.
Since x and y are both odd, $(y - z)$ must also be odd.
Since $(y - z)$ is odd and y is odd, z must be even.
Then, $x + z$ is odd and $x + y$ is even.
Thus, statements I and III can be true.

Practice Questions 2.3

1. (C); 12, 44, 92, and 100 all have two prime factors, while 64 has only one prime factor: 2.

2. (E); 2 is the only even prime number and thus the only even prime factor for all numbers.

3. (C); There are five prime numbers:
37, 41, 43, 47, and 53.

4. (B); The number of distinct prime factors remain the same regardless of whether the number is squared or cubed. Since 28 has two distinct prime factors (2 and 7), 28^2 has two.

5. (C); Since $120h = 2^3 \bullet 3 \bullet 5h = (2 \bullet 3 \bullet 5)^3$
The minimum value of h can be
$3^2 \bullet 5^2 = 225$.

Practice Questions 2.4

1. (D); Multiply all of the fractions by the LCD, 12. The fractions then become 20, 21, and 18.
Therefore, the fractions are arranged as
3/2, 5/3, and 7/4.

2. (D); Multiply all of the choices by the LCD, 100,000. Then, the choices become:
(A) 420,300 (B) 420,340
(C) 420,340 (D) 420,348
(E) 420,347
Choice D is closest to 420,349.

3. (B); The LCM of 4, 12, and 15 is 60.

4. (D); Divide every side by the LCM of 12, 30, and 15, which is 60. Then, $x/5 = y/2 = z/4$.
Since the equation is now in proportional form, we can deduce that $y < z < x$.

5. (D); The LCM of 40 and 60 is 120 minutes, which means that the bells ring simultaneously every 2 hours:
5 times between 8:00 am and 4:00pm.

Practice Questions 2.5

1. (B); Experiment with different number combinations that have a product of 24:

x	y	z
1	2	12
1	3	8
1	4	6
2	3	4

4 is the least possible value of z.

2. (B); For the least possible value of a - b, we need the maximum value of b and the minimum value of a.
Experiment with different number combinations that have a sum of 15:

a	2	3	...	12	13
b	13	12	...	3	2

The answer is 2 - 13 = -11.

3. (B); Experiment with different number combinations that have a product of -6:

a	b	c	d
-3	-2	-1	1
-1	1	2	3

The smallest of these integers is -3.

4. (B); Test the given statements using different number combinations that have a product of 30:

s	2	3	5
t	15	10	6

I: s + t = odd. False.
II: Either s or t is prime. True.
III: s + t + 1 is true for the case of s = 2 and t = 15, but not for s = 3 and t = 10. False.

5. (D); Refer to Section 5.8 for strategies to approach problems involving averages.
By multiplying the average, 5, by 3, we know that the sum of the three integers is 15.
Then, experiment with different number combinations that have a sum of 15, remembering that the greatest product is obtained when the numbers multiplied are closest to each other (Refer to Section 2.5 for Max/Min searches.)

1	2	3	4
2	3	4	5
13	10	8	6

The answer is (4)(5)(6) = 120.

Practice Questions 2.6

1. (E); Experiment with different number combinations using a table like the one below:

X	7	5	6
Y	9	7	5
Sum	79+99 =178	57+77 =134	65+55 =120

Therefore, all three options are possible.

2. (C); The units digit of the powers of 7 can be 7, 9, 3, 1, 7, 9 3, 1, ... and so on.
Therefore, the answer is 5.

3. (C); Let d = 0.XYZ.
Then, 8.74 - (3 + d) = 4 + d
8.74 - 3 - d = 4 + d
1.74 = 2d, and so d = 0.87.

4. (C); Adding up all the units digits yields 14.
Since the hundreds digit is 4, we know that the tens digit needs to be 2_. Thus, A + B + C + 1 should equal 22.
Experiment with different number combinations so that A + B + C = 21.

A	9	9	9
B	9	9	9
C	1 I. No	2 II. No	3 III. Yes

Therefore, III is the only viable answer.

Chapter 2 Test

1. (E); $2x \bullet \frac{y}{2} = xy$.

2. (B); Multiply only the ones digits;
8 • 4 = 32, so the product must end in 2.

3. (D); 47 the only prime number in the list and thus has the least number of factors.

4. (C); There are three positive integer pairs that satisfy the equation 3x + y = 10:
(1, 7), (2, 4), and (3, 1).

5. (A); p(40) = 41 and p(52) = 53.
There are four prime numbers between 41 and 53, inclusively: 41, 43, 47, 53.

6. (C); Since the average of the two numbers is 43, the sum of the numbers is 43 • 2 = 86. The smallest product is obtained when the two most divergent numbers are multiplied (Refer to Section 2.5. for Max/Min searches.)
In this case: 1 + 85 = 86, and 1 • 85 = 85.

7. (A); Since we are given the product of the three digits, 90, find the prime factors:

$90 = 2 \bullet 3^2 \bullet 5$

Therefore, the greatest three-digit number must be 952, and the units digit is 2.

8. (C); Since x, y, and z are measures of the interior angles of a triangle, x + y + z = 180 (See Section 7.1). Since x = y/4 and y = (3/5)z, all of the variables can be expressed in terms of x:

x = x, y = 4x, and z = (5/3)x

Therefore, x + y + z = x + 4x + (5/3)x = 180.

By multiplying each term by the LCD, 3, we get

3x + 12x + 5x = 540

20x = 540, and x = 27

z = 5x/3 = (5/3)(27) = 45.

9. (A); Multiplying each term by the LCD, 4, we get x < 4y < 2x. Since x and y are positive integers when x = 3 and y = 1:

x < 4y < 2x = 3 < 4 < 6,

which is a valid statement.

Therefore, 3 is the smallest possible value for x.

10. (E); Experiment with different number combinations so that x + y = 10. Remember that if we want the least possible value of x - y, we want x to be the smallest possible and y to be the largest possible.

x	1	2	...	8	9
y	9	8	...	2	1
x - y	-8	-6	...	6	8

The smallest value for x - y can be obtained when x = 1 and y = 9.

11. (B); In order for 3,58Q to be divisible by 9, the sum of all its digits must be divisible by 9.

3 + 5 + 8 + Q must equal 18, so Q = 2.

Practice Questions 3.1

1. (C); If (x - s)(x + t) = 0, then x - s = 0 or x + t = 0.

Therefore, x = s or -t.

2. (10/3); Let x = the unknown. Then,

$\frac{x}{100} \bullet 240 = 8$

$x = \frac{8(100)}{240} = \frac{10}{3}$.

3. (E); If $\frac{6}{x} = \frac{y}{14}$, then xy = 84.

The given information is only in regards to xy, not x/y. Therefore, the value of x/y cannot be determined.

4. (E); $x^3 - 4x = (x^2 - 4) = 0$

x(x + 2)(x - 2) = 0.

Therefore, x = 0, -2, and 2.

5. (D); Since x = -1 is one of the solutions, plug -1 back into the original equation to find c:

$(-1)^2 - 2(-1) - c = 0$

1 + 2 - c = 0; Therefore, c = 3.

Then, $x^2 - 2x - 3 = 0$

(x + 1)(x - 3) = 0

The other solution is x = 3.

Practice Questions 3.2

1. (D); Replace 25 with the variable x:

$x^2 - 4x + 3 = (x - 3)(x - h)$.

Therefore, h = 1.

2. (C); First, find the sum of x^2 and y^2: (47)(2) = 94.

Then, use $(x + y)^2$ to solve the problem:

$(x + y)^2 = x^2 + 2xy + y^2 = 94 + 2(3) = 100$.

Thus, $(x + y)^2 = 100$, and since x and y are positive, x + y = 10, and the average is 10/2 = 5.

3. (D); Since $(x - h)^2 = x^2 - 2xh + h^2 = x^2 + 10x + 25$, we know that h = -5.

Then, $(h - 1)^2 - (h + 1)^2 = (-5 - 1)^2 - (-5 + 1)^2$

$= (-6)^2 - (-4)^2$

= 36 - 16

= 20.

4. (D); Since s is the only solution to $x^2 + tx + 100 = 0$ and t > 0, the single solution must come from

$s^2 + ts + 100 = (s + 10)^2 = 0$.

Then, s = -10 Plugging it back into the equation,

$(-10)^2 + t(-10) + 100 = 0$

10t = 200, and t = 20.

Therefore, s + t = -10 + 20 = 10.

5. (D); In order for the equation to be true,

$(x - y)^2 = x^2 - 2xy + y^2 = x^2 - y^2$

$2y^2 - 2xy = 0$

2y(y - x) = 0

2y = 0 or y - x = 0, so y = 0 or y = x.

Practice Questions 3.3

1. (D); Since the question is asking about x in terms of y, through substitution we should first eliminate z:

If 4z = 3y, z = (3/4)y.

Then, (1/4)y = -2x + 3(3/4)y,

and y = -8x + 9y.

Therefore, 8x = 8y and x = y.

2. (C); Changing all expressions to be in terms of a:

a = a, b = a/4, and c = a/3.

The average of a, b, and c, then, is

$\frac{a + a/4 + a/3}{3} = \frac{12a + 3a + 4a}{3(12)} = \frac{19a}{36}$.

3. (B); In order to express z in terms of v, we need to eliminate x and y through substitution:
Since $y = (z/2)$ and $3y + z = 3v$, $3(z/2) + z = 3v$.
Multiplying both sides by the LCD,
$3z + 2z = 6v$, and $z = (6v)/5$.

4. ($\frac{2x}{4y - 3}$); In order to find z in terms of x and y, we need to gather all of the z-related like terms onto one side. By multiplying 4z on both sides,
$4yz = 2x + 3z$, so $4yz - 3z = 2x$
$z(4y - 3) = 2x$
$z = \frac{2x}{4y - 3}$.

5. ($\frac{2 - y}{1 - y}$); By factoring the numerator and denominator,
$\frac{(x - 2)(x - 1)}{(x - 1)(x - 1)} = y$
$\frac{(x - 2)}{(x - 1)} = y$
$xy - y = x - 2$
$x - xy = 2 - y$
$x(1 - y) = 2 - y$
$x = \frac{2 - y}{1 - y}$.

Practice Questions 3.4

1. (B); Since $s - t < 0$, $s < t$. Therefore, only II is true.

2. (A); Since x is a negative fraction, when it is squared, its absolute value becomes smaller. When the exponent is odd, it retains its negative sign. Therefore, when we compare the choices (except B, which is positive), $1/x^3$ is the greatest in absolute value and therefore the smallest as a negative number.

3. (B); Since $x^3 < x^2$, x is either a positive fraction or a negative fraction. However, since $x < x^2$, x must have a negative value $(-1/2 < 1/4)$. $x < x^3$ tells us that x is a negative fraction $(-1/2 < -1/8)$.
Therefore, $-1 < x < 0$.

4. (A); By applying the inequality operational rules of divison,
$$\begin{array}{rcccl} 0.001 & \le & k & \le & 0.1 \\ \div 0.01 & \le & h & \le & 1 \\ \hline 0.001/1 & \le & k/h & \le & 0.1/0.01 \end{array}$$
Therefore, $0.001 \le k/h \le 10$, and the minimum value for k/h is 0.001.

5. (E); By applying the inequality operational rules of subtraction,
$$\begin{array}{c} 7 < b < 9 \\ 4 < a < 6 \\ \hline 7 - 6 < b - a < 9 - 4 \\ 1 < b - a < 5 \end{array}$$
Then, $\frac{1}{5} < \frac{1}{b - a} < \frac{1}{1}$.

6. (C); $6 < \frac{(42 - 3x)}{5}$
$30 < 42 - 3x$
$3x < 12$
$x < 4$, so the greatest integer value would be 3.

7. (D); Since this inequality problem involves a negative number, we must take extra caution when applying the inequality operational rules. While we might be tempted to simply multiply the two inequalities:
$$\begin{array}{c} 4 < a < 6 \\ -2 < b < 2 \\ \hline -8 < ab < 12, \end{array}$$
we must find the smallest possible and largest possible values for each side. The smallest possible value, then, is actually $6(-2) = -12$. Therefore,
$$\begin{array}{c} 4 < a < 6 \\ -2 < b < 2 \\ \hline -12 < ab < 12. \end{array}$$

Practice Questions 3.5

1. (C); Raising a negative number to an even power transform the number to a positive, so
$(-2s^2t^3)^4 = (2s^2t^3)^4$.

2. (C); By factoring and canceling one of the (0.25)'s from the numerator and denominator, we get
$\frac{0.25}{(0.25)(0.25)} = \frac{1}{0.25} = 4$.

3. (D); By changing $(8^k)^2$ into a prime number form,
$[(2)^{3k}]^2 = 2^{6k}$.

4. (A); Factoring each exponent,
$3^{15} - (3^{14} + 3^{13}) = 3^{13}(3^2) - 3^{13}(3 + 1)$
$= 3^{13}(3^2) - 3^{13}(4)$
$= 3^{13}(3^2 - 4)$
$= 3^{13}(9 - 4) = 3^{13}(5)$.

5. (B); $4^y + 4^y + 4^y + 4^y = 4^{y+k}$
$(4^y)(4) = 4^{y+k}$
$(4^y)(4^1) = (4^y)(4^k)$
$k = 1$.

6. (A); $\frac{9^{n+1}}{9^{n+2}} - \frac{9^n}{9^{n+1}} \bullet \frac{(9)}{(9)}$
$\frac{(9^{n+1}) - (9^{n+1})}{9^{n+2}} = 0$.

Practice Questions 3.6

1. ($\sqrt{5}$); Since $(\sqrt{5})(k) = 5$, $k = \frac{5}{\sqrt{5}} \bullet \frac{\sqrt{5}}{\sqrt{5}} = \frac{5\sqrt{5}}{5} = \sqrt{5}$.

2. $(\frac{\sqrt{11}}{30})$; $\sqrt{\frac{1}{25} - \frac{1}{36}} = \sqrt{\frac{36}{900} - \frac{25}{900}} = \sqrt{\frac{11}{900}} = \frac{\sqrt{11}}{30}$.

3. (E); If $x^2 = a^2$, $x = \pm a$.
Then, $xa < 0$ and $x + a < 0$ as well.

4. (5); Isolate the radical term on one side and then square both sides.
$1 + 2 = \sqrt{2x - 1}$
$9 = 2x - 1$
$2x = 10$, and $x = 5$.

5. (A); Since both $\sqrt{a + b}$ and $\sqrt{a} + \sqrt{b}$ are positive, we could square both terms and compare them:
$(\sqrt{a + b})^2 = a + b$
$(\sqrt{a} + \sqrt{b})^2 = a + 2\sqrt{ab} + b$
$a + 2\sqrt{ab} + b > a + b$, and subtracting a larger number from a smaller number will result in a negative.

6. (C); If $a > b$, then $a - b > 0$.
Then, $\sqrt{(a - b)^2}$ and $a - b$ are both positive.

Practice Questions 3.7

1. (A); Putting each equation in the form of $y = mx + n$ where m is the slope, we find that $m = -2/3$ for both lines, which means they are parallel. Therefore, there is no common solution.

$2x + 3y = 6$ | $4x + 6y = 6$
$y = (-2/3)x + 2$ | $y = (-2/3)x + 1$

2. (D); The point where a line crosses the x-axis is the x-intercept, which is found by plugging in 0 for y.
$3x - 2(0) = 12$, and $x = 4$.

3. (E); The graph $y = -1$ is a horizontal line along the x-axis because it is not affected by the values of x. Therefore, it does not touch quadrants I and II.

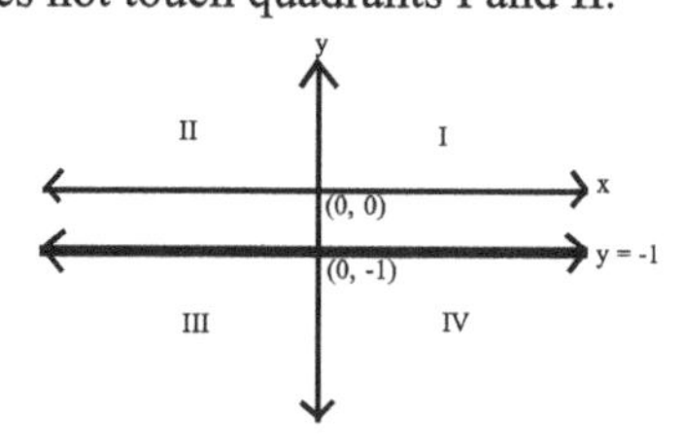

4. (A); Find the slopes of the two lines by changing the equations into standard form, $y = mx + n$ where m is the slope:

$2x - 5y = 3$ | $10y - 4x = 3$
$y = (2/5)x + 3$ | $y = (2/5)x + 3/10$

Both slopes are the same, making the lines parallel with no common solution.

5. (D); In order to find a function, draw a vertical line through any graph to see if there is more than one point of intersection, which makes it not a function. Only answer (D) has one intersection for each vertical line drawn, so it is the only function.

6. (C); Plug the coordinates into the slope formula:
slope $= \frac{y^2 - y^1}{x^2 - x^1} = \frac{-4}{3}$
$\frac{-1 - t}{-2 - (-5)} = \frac{-4}{3}$
$\frac{-1 - t}{3} = \frac{-4}{3}$
$-1 - t = -4$, so $t = 3$.

7. (D); The only difference between the two equations is the slope, m, which goes from 1 to -1/2. Therefore, the new graph will go in the opposite direction with a less-steep slope, but its x- and y-intercepts remain the same.

8. (E); $x - 2y = 5$ in standard form is $y = (1/2)x - 5$, which means that the slope is 1/2. A perpendicular line is the negative reciprocal of the slope: -2. Therefore, line T is the graph $y = -2x + n$. To find n, plug in the coordinates (-1, 2):
$y = -2x + n$
$2 = 2(-1) + n$
$2 = 2 + n$
$n = 0$
Therefore, line T's equation is $y = 2x$.
When $x = 0$, $y = 0$, and when $y = 0$, $x = 0$.
The x- and y-intercepts are same at the origin.

Practice Questions 3.8

1. (B); $g(2) = -f(2 - 1) + 1 = -f(1) - 1$
Since $f(1) = 0$, $g(2) = 0 - 1 = -1$.

2. (D); Since a, the x-coordinate, is bound by the graph f(x) and the x-axis, the maximum x-value is where f(x) crosses the x-axis on the positive side, which is 1.

3. (B); Since $f(x) = x^2 + 1$ and $f\sqrt{y + 1} = 1$,
$f(x) = (\sqrt{y + 1})^2 + 1$.
Then, $y + 1 + 1 = 1$, so $y = -1$.

4. (B); According to the graph, $k(0) = t = 1$.
From the same graph, then, $k(1) = 0$.

5. (A); $g = k(-x + 1) = k[-(x - 1)]$.
Therefore, the graph is shifted to the right and reflected downwards across the y-axis.

Practice Questions 3.9

1. (-1); Since $(x - 1)^2 = (x + 3)^2$,
   $x - 1 = x + 3$ or $x - 1 = -x - 3$
   Not possible $2x = -2$, so $x = -1$.
   Therefore, the only possible solution for x is -1.

2. (B); Substitute x into the first equation:
   $y^2 + 4 + 3y = 4$
   $y^2 + 3y = 0$
   $y(y + 3) = 0$. $y = 0$ or -3.
   Therefore, -3 is a possible value for y.

3. (C); If $x + 3$ is a factor of $x^2 - tx - 12$, then $x = -3$ is a solution that we could use to plug into the equation:
   $(-3)^2 - (-3)t - 12 = 0$
   $9 + 3t - 12 = 0$
   $3t = t$, so $t = 1$.

4. (B); If $x^2 + 5x - 6 = 0$ and $x < 0$,
   then $(x - 1)(x + 6) = 0$. Then, $x = 1$ or -6.
   Since x is negative, $x = -6$. Plugging it into the choices:
   $x^2 - 5x - 6 = (-6)^2 - 5(-6) - 6$
   $= 36 + 30 - 6 = 60, \neq 0$.

5. (E); If $(x - k)^2 = 9$, then $x - k = +3$ or -3.
   Then, $x = k + 3$ or $k - 3$.
   The sum of the values of x would be
   $k + 3 + k - 3 = 2k$.

6. (D); By plugging in the x values and checking the corresponding f(x) (or y) values,
   (1, 0): $1 = -0^2 + 1$, so $1 = 1$. $f(x) = -x^2 + 1$.

7. (B); Since the coefficient a is negative and b is positive, $f(x) = -(-a)(x^2 - b/a) + c$. Therefore, the graph wouldopen downwards and be shifted to the right.

8. (2); Since the graph $y = 2x^2 - 2$ intersects line k at (t, 4) and (p, 0), the equation $y = 2x^2 + 2$ should hold true at both points.
   $4 = 2t^2 + 2$ $0 = 2p^2 + 2$
   $t = +1$ or -1 $p = +1$ or -1
   Since the slope of line k is $\frac{y^2 - y^1}{x^2 - x^1} = \frac{0 - 4}{p - t}$,
   the maximum value can be obtained when $p = -1$ and $t = 1$. Therefore, the slope is $-4/(-1 - 1) = 2$.

9. (10/3); By plugging the point (3, t) into both equations,
   $t = (1/3)(3) + 2$ $t = (-1/9)(3) + k$
   $t = 1 + 2 = 3$ $t = -1/3 + k$
   Equating the two, $3 = -1/3 + k$, so $k = 10/3$.

10. (1); The two graphs are parabolas facing upwards and downwards towards each other. Equating the two equations,
   $2x^2 = 4c^2 - 2x^2$
   $4x^2 = 4c^2$
   $x = +c$ or $-c$
   Since the x-coordinates of the two intersections are +c and -c, the distance between them is 2c.
   From the given information, $2c = 2$, so $c = 1$.

11. (3); g(x) is f(x) shifted left by m and upwards by n, and $m = 2$ and $n = 1$. Then, $m + n = 3$.

Practice Questions 3.10

1. (3); When $a > 0$, $a + 5 > 0$. Then $a + 5 = 7$ and $a = 2$.
   When $b < 0$, $b - 2 < 0$. Then $-(b - 2) = -b + 2 = 3$ and $b = -1$.
   Therefore, $a - b = 2 - (-1) = 3$.

2. (B); The function is satisfied when the difference between the absolute values of y and x is 1.
   $|3| - |-2| = 1$.

3. (E); If $650 < T < 750$,
   $-50 < T - 700 < 50$
   So $|T - 700| < 50$.

4. (B); $|-2 - x| < 5$, and $|2 + x| < 5$
   Therefore, $-5 < 2 + x < 5$, and $-7 < x < 3$.
   $x = -5$ is the only choice fitting the boundaries.

5. (C); According to the piece-wise definition, we want to consider the cases in which $x + 2 = 0$, and $x = -2$.
   When $x \geq -2$, h(x) is positive $= x + 2$.
   When $x < -2$, h(x) is negative $= -(x + 2) = -x - 2$.

6. ($x > -3/2$); Since $-3 < x < -1$, $x + 3 > -x$.
   $2x > -3$
   $x > -3/2$.

7. ($x > 5$ or $x < 0$); We need to transform the equation to its piece-wise form:
   When $2x > 5$, then $2x - 5 > 0$. $x > 5$.
   When $2x < 5$, then $-(2x - 5) > 5$.
   $-2x + 5 > 5$
   $-2x > 0$, so $x < 0$.
   Therefore, $x > 5$ or $x < 0$.
   As a short-cut, $2x - 5 > 5$ or $2x - 5 < -5$.
   Then, $x > 5$ or $x < 0$.

Chapter 3 Test

1. (B); Find the product of $(x + y)(1/x + 1/y)$ using the FOIL method:
   $1 + x/y + y/x + 1 = 2 + x/y + y/x = 2 + 5 = 7$.

2. (E); Since $x^2 - sx + 8$ factors into $(x + 1)(x + t)$, we can figure out that $t = 8$.
   Then, $(x + 1)(x + 8) = x^2 + sx + 8$, so $s = 9$.

3. (E); Since the question is asking about the value of z, express all three angles in terms of z. Then, equate the sum to 180, which is the sum of the interior angles of a triangle.

$x = \frac{y}{3} = \frac{3z}{5 \bullet 3} = \frac{z}{5}$, $y = \frac{3z}{5}$, and $z = z$

$x + y + z = 180$

$\frac{z}{5} + \frac{3z}{5} + \frac{5z}{5} = 180$

$\frac{9z}{5} = 180$, so $9z = 900$ and $z = 100$.

4. (D); According to the operational rules of inequalities in subtraction,

$$\begin{array}{c} 7 < y < 10 \\ 2 < x < 5 \\ \hline 7 - 5 < y - x < 10 - 2 \\ 2 < y - x < 8 \end{array}$$

Then, $\frac{1}{2} > \frac{1}{y - x} > \frac{1}{8}$.

5. (A); By multiplying the three sides of the inequality by the LCD, 4, we get $x < 4y < 2x$. Since x and y are both positive integers, the smallest possible value for x must be at least 3 so that $y = 1$.

Thus, $3 < 4 < 6$ is a viable solution, where $x = 3$.

6. (C); Remember the exponential rule: $(x^a)^b = x^{ab}$.

$(4^n)^3 \neq 12^n$.

7. (B); Since the graph of $x = -1$ is a vertical line while the graph of $y = 3$ is a horizontal line, there is only one intersection and thus one solution: (-1, 3).

8. (D); $(a - b)^2 = a^2 + b^2 - 2ab$, which is $18 - 2(6) = 6$.

9. (E); Plugging in the answer choices, $x = 4$ and $y = -4$ is the only choice in which $x^2 + y^2 > 25$:

$4^2 + (-4)^2 = 16 + 16 = 32 > 25$.

10. (D); Since one of the functions is an absolute value, we need to consider both postive and negative cases:

$x + 1 = 1$	$-(x + 1) = 1$
$x = 0$	$-x - 1 = 1$
	$x = -2$

Therefore, the graphs intersect at $x = 0$ and -2.

11. (A); Changing both equations to the point-slope form of $y = mx + b$, we can see that $m = -1$ for both equations, making the lines parallel and without any common solutions.

$x + y = 10$	$x + y = 15$
$y = -x + 10$	$y = -x + 15$
$m = -1$	$m = -1$

12. (E); Isolate the radical on one side, then square both sides to eliminate it:

$2 - \sqrt{3x - 5} = -2$

$4 = \sqrt{3x - 5}$

$16 = 3x - 5$

$3x = 21$

$x = 7$.

13. (D); $y = 2(x - 1)^2$ is $g(x) = -x^2$ flipped to the opposite direction (facing upwards because of +x), shifted to the right by 1 unit (because of -1), and with a steeper slope (because of 2).

14. (D); Gathering all of the x-related like terms, and then factoring in terms of x:

$ax - bx = ay - by$

$x(a - b) = y(a - b)$

$x = \frac{y(a - b)}{(a - b)} = y$.

Practice Questions 4.1

1. (E); $3 @ (11 @ 2) = 3 @ \frac{(11 - 2)}{3} = 3 @ 3 = \frac{3 - 3}{3} = 0$.

2. (B); Since x^ is the greatest integer less than x, (3.7)^ = 3 and (-3)^ = -4.

Thus, 3 + (-4) = -1.

3. (E); Since x*** = $x - x^2$, $x^2 - x + x - x^2 = 0$.

Since (1)*** = $(1) - (1)^2 = 0$, the answer is E.

4. (C); I: x # y = y # x is false.

$(x + y)(x - y) = x^2 - y^2$

$(y + x)(y - x) = y^2 - x^2$

II: x # 0 = 0 # x = x^2 is false.

$x^2 = -x^2$

III: x # -y = x # y is true.

$(x - y)(x + y) = x^2 - y^2$

$(x + y)(x - y) = x^2 - y^2$.

Therefore, only III is true.

5. (B); According to the definition, 4!b = 4b + (the greatest prime factor of 4b) = 55.

When b = 11, the greatest prime factor of 4(11), or 44, is 11.

Then, 44 + 11 = 55.

6. (C); By the Bird's-Eye View (Don't Calculate but Write Out) approach:

(>97) = (total # of prime numbers between 2 and 97)

(>105) - 2 = (total # of prime numbers between 2 and 105) - 2.

Therefore, we are looking for (total # of prime numbers between 97 and 105) - 2:

There are two prime numbers between 97 and 105: 101 and 103.

2 - 2 = 0, so (>97) = (>105) - 2.

Practice Questions 4.2

1. (D); When the number of cards given to friends (6x) is subtracted from the total (s), the remaining number of cards is s - 6x.

2. (A); Using the division rule outlined in the formal approach to this type of problem:
$n = 6Q + 2$
Then, $3n + 1 = 3(6Q + 2) + 1$
$= 18Q + 6 + 1$
$= 18Q + 7$.
If $18Q + 7$ is divided by 6, the answer is $3Q + 1$ and the remainder is 1.

3. (E); By experimenting with the numbers between 110 and 120, we find that 118 is divisible by 7 with a remainder of 6, and divisible by 6 with a remainder of 4.
Therefore, there were 118 people in the theatre.

4. (A); When k is divided by 6, the remainder is 2:
$k = 6Q_1 + 2$
When h is divided by 6, the remainder is 4:
$h = 6Q_2 + 4$
Therefore, $k + h = 6Q_1 + 6Q_2 + 6$.
Then, k + h divided by 6 = $Q_1 + Q_2 + 1$, and the remainder is 0.

5. (C); Converting this division formula into an equation, the number becomes 6Q + 5.
$\frac{6Q + 5}{3} = 2Q + 1$ with a remainder of 2.

Practice Questions 4.3

1. (A); Since the sequence repeats every four terms, dividing 205 by 4 gives us the remainder, 1. Therefore, the 205th term will be the first term of the sequence: 2.

2. (B); The fourth number and numbers after it will be:
24, 41, 74, 139, 254.
The 8th number, 254, is the first to exceed 200.

3. (D); Upon close scrutiny, the general sum is represented by (5/2)(n)(n + 1).
For example, for 10 + 5 = (5/2)(2)(3),
the sum is (5/2)(2 terms)(2 + 1)
= (5/2)(2)(3).

4. (D); The movement of the yo-yo is depicted below:

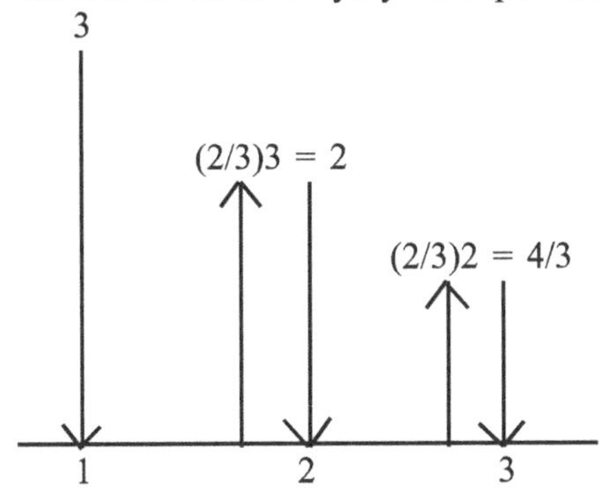

The total vertical distance traveled by the yo-yo will be:
3 + 2 + 2 + 4/3 + 4/3 = 7 + 8/3 = 29/3.

5. (D); According to the formula described:
$1092 = 1/2(a_{n+1} - 3)$
So, $2184 = a_{n+1} - 3$ and $a_{n+1} = 2187$.
2187 is the 7th term of the sequence, so n = 7.
3, 9, 27, 81, 243, 729, 2187

6. (D); Working backwards according to the given formula $k_n = k_{n-1} - 3$,
$k_4 = 25 = k_3 - 3$
$k_3 = 28 = k_2 - 3$
$k_2 = 31 = k_1 - 3$
$k_1 = 34$.

Practice Questions 4.4

1. (B); The original statement is the same as the contrapositive statement.

2. (C); Since the statement "must" be true, the answer must be absolutely true. While we cannot be sure about the other statements, statement C must be true.

3. (C); In order to have the minimum number of moves, we want to advance the most possible spaces at a time per move. Dividing 23 spaces by 4 spaces/move = 5 moves, with a remainder of 3. The 6th and last move, then, is for 3 spaces. Therefore, 6 is the minimum number of moves required to advance 23 spaces.

4. (B); We can summarize the data using the table below:

	New	Used	
Red	5 - 3 = 2	15 - 12 = 3	(1/4) • 20 = 5
Green	15 - 12 = **3**	12	20 - 5 = 15
	(1/4) • 20 = 5	20 - 5 = 15	

5. (C); The following table summarizes each scenario in which Player A removes 3 cards at her first turn in order to secure her win.

A	B	C	A
3	5	1	1
3	4	1	2
3	3	1	3
3	2	1	4
3	1	1	5

Therefore, if Player C removes only 1 card at her turn, no matter how many cards Player B removes, there is no way that Player A can lose.

6. (B); The multiples of 3 start from 3 and end with 999, but they are excluded in this case. Therefore, there are (333 - 1 = 332) multiples of 3, excluding 3 and 999.

Practice Questions 4.5

1. (E); The maximum possible cases from tossing a coin (heads or tails) four times is (2)(2)(2)(2) = 16. The only possible case in which the sum of the four tosses is less than four points is when all four coins land on heads (1 point each), which is just one case out of the 16 possible: 1/16.

The probability that the sum of the four tosses is at least 5, then, is 1 - 1/16 = 15/16.

2. (C); The probability that the first card drawn is a 10 is $\frac{\text{four 10's in a deck}}{\text{52 cards total}} = \frac{1}{13}$.

3. (E); If the first two digits selected are 8 and 6, the remaining possible numbers for the last two digits are 0, 2, 4, 1, 3, 5, 7, and 9. That means there are eight possibilities for the third digit, and seven possibilities for the fourth: (8)(7) = 56 total possibilies.

The events favored are 3 even numbers for the ones digit and the remaining numbers for the tens digit.

Therefore, (3)(7) = 21, and the chances that the four digit number is even is 21/56 = 3/8.

4. (B); While switch 1 can be either on or off, switch two must always be in the opposite state. Switches 3 and 4 can be either on or off. Therefore, there are 2 possibilities for switch 1, one possibility for switch 2, and two possibilities for both switches 3 and 4.

(2)(1)(2)(2) = 8 different combinations.

We could also write out all of the possible arrangements:

1	2	3	4
on	off	off	off
on	off	on	off
on	off	on	on
on	off	off	on
off	on	off	off
off	on	on	off
off	on	on	on
off	on	off	on

5. (B); This question is essentially asking how many different ways can the three letters, b, c, and d, can be arranged (one-time choice and arrangement: permutation). Therefore, there are 3 possibilities for the first spot, 2 for the second, and 1 for the third.

3! = (3)(2)(1) = 6 possible arrangements.

We could also write out all of the possible arrangements:

bcd, bdc, cbd, cdb, dbc, and dcb.

6. (B); The probability that the first marble chosen from the jar is red is 3/8.

The probability that the second marble chosen from the jar is 2/7.

Therefore, the probability that two marbles randomly chosen without replacement is

$\frac{3}{8} \bullet \frac{2}{7} = \frac{6}{56} = \frac{3}{28}$.

7. (E); This is another permutation problem, as it deals with one-time choice and arrangement. However, this time there are only three spots available, which means we will use just n(n - 1)(n - 2). There are 10 contestants eligible for the 1st prize, 9 for the 2nd prize, and 8 for the third prize.

Therefore, (10)(9)(8) = 720 possible ways.

8. (A); We are looking for any 3-person combination, regardless of order.

$_{10}C_3 = \frac{10 \bullet 9 \bullet 8}{3 \bullet 2} = 120.$

9. (39/50); This is a standard probability problem in which we are looking for at least 1 case (Case 1 or 2). The probability that a randomly selected employee is a female or Caucasian is

p(female) + p(Caucasian) - p(both female and Caucasian)

$= \frac{58}{100} + \frac{38}{100} - \frac{18}{100}$

$= \frac{78}{100} = \frac{39}{50}$.

10. (0.6); This is a simultaneous/independent probability problem in that Hank and Babe's batting performances do not influence each others'.

The probability that at least one player will hit a home run in today's game is (0.3)(0.2) = 0.6.

11. (90);Since the five numbers constitute a five-digit even number, the first digit cannot be zero. Furthermore, the last digit must be one of the three even numbers. The middle three numbers can then be arranged through a permutation from the remaining three numbers:

$5 \bullet 3 \bullet {}_3P_3 = 15 \bullet 3! = 90.$

Practice Questions 4.6

1. (3); Arranging the scores in ascending order, we get 2, 2, 2, 2, 3, 3, 3, 3, 3, 4, 4, 4, 4, 4, x.

Regardless of whatever value x assumes, the median or 8th number will be 3.

2. (4); In order for the mode to be 4, the score 4 must be the most frequent score. Therefore, x must be 4.

3. (2); If the mean and median are equal, x must be 2 so that all scores are equally distributed with the frequency of 5. If x = 2, both the mean and median would be 3.

4. (4); Let T = the total number of times the two additional students took the SAT. The total number of students surveyed is 1 + 6 + 8 + 8 + 2 = 25. The average number of times each student has taken the SAT, then, is the weighted average of the total:

$$\text{Average} = \frac{(0)1 + (1)6 + (2)8 + 3(8) + T}{25}$$

Arranging the number of tests from least to most times taken, since the median number is the number of times the middle student has taken the SAT, it would be the 13th student's result: 2 times.

$$\text{Then, } 2 = \frac{46 + T}{25}$$

46 + T = 50, so T = 4 times.

5. (650); The median of 10 scores is the average of the middle two (5th and 6th) scores. The median is 650.

6. (30); The new scores are 640, 650, 650, 680, 680, 680, 770, 770, 780, and 800.

The new mean score is 710.
The new mode score is 680.
710 - 680 = 30 points.

7. (7); To find the mean number of college applications submitted by all 60 students, divide the total number of applications submitted from all three classes by the total number of students.
First, find the total number of applications submitted:

Class A: (20 students)(8 apps/student) = 160
Class B: (30)(6) = 180
Class C: (10)(8) = 80
Therefore, there were 420 applications total submitted by the students in Classes A, B, and C.

$$\text{Then, the mean} = \frac{420 \text{ apps}}{60} = 7 \text{ apps/student.}$$

8. (D); To find the median for the 60 students, we want to find the average of the number of college applications submitted by students #30 and #31. In this case, both students #30 and #31 are in Class B, which means, according to the table, the median is 8.

9. (3.5); $\frac{4 + m + 5 + 9 + 3 + n}{6} = 4$

21 + m + n = 24
m + n = 3
Since m > n and they are both non-negative integers, we can deduce that m = 2 and n = 1.
The sequence, then, is 1, 2, 3, 4, 5, and 9.
The median is the average of the third and fourth numbers, 3 and 4, which is 3.5.

Practice Questions 4.7

1. (340); According to the table, 34% of the total number of people surveyed are against abortion.

(0.34)(1,000) = 340.

2. (300); If we let x = the number of men,

0.5x + 0.6(1,000 - x) = 0.53(1,000)
0.5x + 6,000 - 6x = 5,300
x = 700 men
Therefore, there are 1,000 - 700 = 300 women.

3. (A); The student-class ratio is represented by the slope of the line connecting the origin to each point on the graph. Then, the class-student ratio is the reciprocal of the slope of each line.

Since the slope is 1994 is the smallest, it represents the smallest student-class ratio and thus the highest class-student ratio.

4. (5); The number of students taking Geometry is 30% of 500, or 150 students. Then, the number of students withdrawn from Geometry is 10% of 150, or 15.
To find the number of stdudents taking Calculus,

x + 3x + 50 + 30 = 100
4x + 80 = 100
4x = 20, so x = 5.
5% of 500 is 25 students.

Then, the number of students failing Calculus is 40% of 25, or 10.

15 - 5 = 5 students.

5. (15%); We found from the previous problem that x = 5% of students taking Calculus. Then, 3x = 3(5) = 15% of 500 students, or 75 students are taking Algebra II. Then, 20% of 75, or 15 students are failing Algebra II.

6. (16%); Make a chart depicting the number of students failing in each of the math classes:

	Number of Students
Algebra I	10% of 250 = 25
Geometry	20% of 150 = 30
Algebra II	20% of 75 = 15
Calculus	40% of 25 = 10
	80 students total

Then the overall percentage of students failing across all the math classes is 80/500 or 16%.

7. (B); While John's family's automobile expenses were 7% of its total income in both 2009 and 2010, since the total income in 2010 was greater than the total income in 2009, the automobile expenses in 2010 were thus greater than the automobile expenses in 2009.

8. (C); Note that the graph shows the growth rate of the GDP, as opposed to the actual number!
(A) The GDP growth rate, not the actual GDP, was highest in 2004.
(B) The actual GDP is not necessarily higher.
(C) The actual GDP is lower in 2008 than in 2007 because the year recorderd a negative growth rate.
(D) Though the point in 2010 is higher than the point in 2007, this is not indicative of a higher GDP.

9. ($940); For Jason, x% = 100 - 15 - 25 - 30 = 20%.
20% of $2,000 = $400.
For Steph, y% + z% = 100 - 20 - 10 - 25 = 45%.
45% of $1,200 = $540.
Therefore, x + y + z = 400 + 540 = $940.

10. (B); (A) Savings increased by 80%, from $5,000 to $9,000. True.
(B) The savings/income ratio would be $5,000/$40,000, $7,000/$45,000, and $9,000/$50,000. False.
(C) The percent increase in 2000-2005 is $2,000/$5,000, while the percent increase in 2005-2010 is $2,000/$7,000.
(D) There is a $2,000 increase in income between both 2000-2005 and 2005-2010.

Chapter 4 Test

1. (E); $< 3.2\# > = 4$ and $< -2.5\# > = -2$.
Therefore, 4 - (-2) = 6.

2. (C); When the division formula is converted into an equation, it is
n = 5Q + 1.
2n = 10Q + 2, and when it is divided by 5, the remainder is 2.

3. (C); The sequence of the numbers are as follows:
5, 6, 11, 17, 28, 45, 73, 118...
The 8th number, 118, is the first to exceed 100.

4. (A); Let n = the number of computers. Then, when n is divided by 4, the remainder is 3, and when n is divided by 3, the remainder is 2. 11 is the only number that satisfies this condition.

5. (E); The sequence of numbers is as follows:
First term: 2
Second term: $2^3 + 1 = 9$
Third term: $9^3 + 1 = 730$.

6. (B); According to the definiton given,
(1)(9) - (1)(2) = 9 - 2 = 7.
The greatest prime factor of 7 is 7.
(3)(33) - (3)(17) - 99 - 51 = 48.
The greatest prime factor of 48 is 3.
7 - 3 = 4.

7. (E); The only true statement is that since the average of all 11 integers is y, the sum of all the integers is 11y.

8. (B); 35 is the 5th multiple of 7, while 693 is the 99th multiple of 7. Therefore, using the formula for counting numbers in a sequence while excluding the border numbers, 99 - 5 - 1 = 93.

9. (B); According to the question, the sum of the first k terms is $(1/2)a_{k+1} - 2) = 160$.
Then, $(1/2)a_{k+1} = 162$, and $a_{k+1} = 324$.
Thus, the sequence is as follows:
4, 12, 36, 108, 324, ...
Since k + 1 = 324, k is the 5th term.
Therefore, k is the 4th term.

10. (B); A one-time choice and arrangement of the 3 letters and 3 numbers makes it a permutation problem. Since the orders of the letters and numbers occur simultaneously/independently of each other, we multiply the two permutations:
xyz: (3)(2)(1) = 6 combinations
123: (3)(2)(1) = 6 combinations
(6)(6) = 36 total combinations.

11. (E); In these situations, the probability for the first student is always 1, since it does not matter in which month that student was born. The probability that the second student was born in a different month is 11/12. Since these events occur simultaneously/independently of each other, we multiply them:
(1)(11/12) = 11/12.

12. (C); Statement III is true, since it is the contrapositive of the original statement.

13. (B); Let x = the number of students in the 10th grade, and let y = the number of students in the 11th grade.
Then, 0.3x + 0.6y = 0.5(x + y)
0.3x + 0.6y = 0.5x + 0.5y
0.2x = 0.1y
y = 2x, so y > x.
Therefore, there were more students in the 11th grade than in the 10th grade.

14. (E); All statements are true.

15. (D); The US owes $4.5 trillion out of $14.13 trillion to foreign countries, which is about 31.46% of the total US debt. Therefore, the debt owed to China would be about 8% of the total.

Practice Questions 5.1

1. 2x - 3x = 5.

2. (1/2)x + (2x - 0.3x) = x - 3.

3. Let x = John's current age.
Then, 4(x - 5) = 3(x + 2) - 3.

4. $x + \frac{1}{x} = \frac{2x + (1/2)x}{x + 5}$.

Practice Questions 5.2

1. (4 worms); Let x = the number of worms in the orange.
Then, 4x = the number of worms in the apple.
$4x - 5 = 11$
$4x = 16$, so $x = 4$.

2. (15); $x + 6 = y + 7$
$x + 6 + 12 = y + 7 + 12$
$x + 18 = y + 19$
$(y + 19) - (y + 4) = 15$.

3. (12 pounds); Let x = the weight of the stone
Then, $x + (1/4)x = 15$
$4x + x = 60$
$5x = 60$
$x = 12$.

4. (6x + 13 dollars); Since Irene won 3x dollars,
Jack = Irene + 4 = 3x + 4
Sarah = Jack + 5 = (3x + 4) + 5 = 3x + 9
Therefore, Sarah + Jack = 3x + 9 + 3x + 4
= 6x + 13.

5. (3:1); Let W = Wendy's age, and M = Michael's age.
Then, $\frac{W + M}{2} = W - M$
$W + M = 2W - 2M$
$W = 3M$
$W/M = 3/1$.

Practice Questions 5.3

1. (6); Let x = the first even integer.
Then, x + 2 = the second even integer.
$x + x + 2 = 10$
$2x = 8$, so $x = 4$.
Therefore, the larger integer is 4 + 2 = 6.

2. (A); Statement I is not necessarily true because the three consecutive integers could be 1, 2, and 3, none of which are evenly divisible by 4.
Statement II is not necessarily true because the digits could be anything other than 5: again, 1, 2, and 3.
Statement III is not necessarily true because if the sequence starts with an odd integer, only one of the numbers would be even and thus divisible by 2: again, 1, 2, and 3.

3. (B); If the product of these numbers is 0, one of them must be 0. Then, the numbers at their maximum values are 0, 1, 2, 3, and 4, the sum of which is 10.

4. (B); If x = the first integer, x + 1 = the second integer, and x + 2 = the third integer,
$x + x + 1 + x + 2 = 3x + 3$
which is divisible by 3.

5. (C); Let x = the first integer, x + 2 = the second even intger, x + 4 = the third even integer, and x + 6 = the last even integer.
Then, $x + x + 2 + x + 4 + x + 6 = 72$
$4x + 12 = 72$
$4x = 60$
$x = 15$.
Therefore, the there integers are 15, 17, 19, and 21, and the average of the smaller three is
$\frac{15 + 17 + 19}{3} = 17$.

Practice Questions 5.4

1. (30A + 3B); The number with tens digit A and units digit U is represented by 10A + B. Then, the product of 3 and (10A + B) is 30A + B.

2. (D); The two-digit number is 10t + v, while the new three-digit number is 100t + 10v + m.
$10(10t + v) - (100t + 10v + m)$
$= 100t + 10v - 100t - 10v - m$
$= -m$.

3. (D); Let t = the tens digit and u = the units digit.
$(10t + u) - (10u + t) = 9t - 9u$.
Therefore, all answers must be divisible by 9, which 64 is not.

4. (E); The numbers added can be expressed as
$100A + 10B + A + 100B + 10A + B$
$= 111A + 111B$.
Therefore, the answer must be divisible by 111, which 999 is.

Practice Questions 5.5

1. (9); Let x = how old the child will be.
Then, 27 + x = how old the woman will be.
When the woman's age is 4 times the child's,
$4x = 27 + x$
$3x = 27$, so $x = 9$.

2. (D); Let x = Mary's age today.
Then, x - 9 = Mary's age 9 years ago.
Since Jack is now Y years old and his age is twice of x - 9,
$Y = 2(x - 9)$
$Y = 2x - 18$
$2x = Y + 18$
$x = \frac{Y + 18}{2}$

3. (11); Benjamin and Jeremiah's ages are shown below:

	Now	In 5 years
Benjamin	x	x + 5
Jeremiah	x - 8	x - 8 + 5 = x - 3

Then, $x + 5 = 2(x - 3)$
$x + 5 = 2x - 6$
$x = 11$.

4. (C); Let x = the number of years it takes for Sandy to become twice as old as Sharon.

Sandy	T	T + x
Sharon	T/4	(T/4) + x

Then, $T + x = 2((T/4) + x)$

$T + x = T/2 + 2x$

$x = T/2$.

5. (10 years); Marge and Homer's ages are shown below:

	-20 yrs	-10 yrs	Current	+5 yrs
Marge		25	35	40
Homer	20		40	45

Homer's age 5 years from now is 45, while Marge's current age is 35.

Thus, 45 - 35 = 10 years.

Practice Questions 5.6

1. (1/5 minute); Let x = unknown.

$$\frac{50 \text{ miles}}{1 \text{ second}} = \frac{600 \text{ miles}}{x \text{ seconds}}$$

50x = 600, so x = 12 seconds = 1/5 minute.

2. (18 meters); Let x = unknown. Since the ratio of the height of an object to the length of its shadow is constant, $\frac{x}{72} = \frac{3}{12}$, and x = 18 meters.

3. (E); Let G = the unknown. Through the proportion formula, $\frac{x \text{ pounds}}{n \text{ grams}} = \frac{p \text{ pounds}}{G \text{ grams}}$

$xG = np$

$G = np/x$.

4. (1/16 pints); Let x = the unknown. Use the proportion formula, keeping in mind the unit conversion:

1 quart = 2 pints

$$\frac{2 \text{ pints punch}}{1/2 \text{ pints sprite}} = \frac{1/4 \text{ pints punch}}{x \text{ pints sprite}}$$

$2x = (1/4)(1/2)$

x = 1/16 pints of sprite.

5. ($750); Let x = the unknown. Since the cost of the carpet is proportional to its area, use the proportion:

$$\frac{(8)(4) \text{ feet}}{\$500} = \frac{(24)(2) \text{ feet}}{\$x}$$

$$\frac{32}{500} = \frac{48}{x}$$

x = 750.

Practice Questions 5.7

1. (3/2); 3/5 of the marbles are red, and 2/5 of the marbles are not red. Therefore, the ratio is

$$\frac{3/5}{2/5} = 3/2 \text{ .}$$

2. ($\frac{y}{x+y+z}$ cups); By adding up all of the ingredients in the whole recipe is comprised of x + y + z cups.

3. (3 games); The sum of the ratio is 5 + 4 = 9, which goes thrice into 27. Multiplying each number in the ratio by 3, then, will yield the number of games won and lost:

5 wins • 3 = 15 wins

4 losses • 3 = 12 losses

The team has won 15 games and lost 12;

15 - 12 = 3.

4. (28:25); The ratio can be figured out as follows:

m	:	n	:	p
5	:	7		
4	:			5
20	:	28	:	25

Therefore, the ratio between n and p is 28:25.

5. (81); Find the ratio of children and adolescents to the total number of people at the park:

$$\frac{\text{children + adolescents}}{\text{total people at the park}} = \frac{15 + 12}{3 + 15 + 12 + 20} = \frac{27}{50}$$

Therefore, the total number of children and adolescents in the park:

150 people total • $\frac{27}{50}$ = 81.

Practice Questions 5.8

1. (14); The sum of the three numbers is 10 • 3 = 30.
Then, 16 + x = 30, so x = 14.

2. (60); The sum of the three numbers is 20 • 3 = 60.

3. (65); The sum of the four numbers is 17 • 4 = 68.
Then, 3 + x = 68, so x = 65.

4. (18); The sum of the three numbers is 10 • 3 = 30, while if y is added, the new total will be 30 + y.
Then, 30 + y = 48, so y = 18.

5. (84); The total of Peter's Math and English scores is 90 • 2 = 180, while the total of Peter's Science, Spanish, and P.E. scores is 80 • 3 = 240. Therefore, the total score across all five subjects is 180 + 240 = 420.

420/5 subjects = 84.

Practice Questions 5.9a

1. (100); $\frac{25}{100} \bullet 300 = \frac{75}{100} \bullet x$

$$\frac{300}{4} = \frac{3x}{4}$$

x = 100.

2. (80%); $X = \frac{150}{100} \bullet Y$, so $Y = x \bullet \frac{100}{150} = \frac{4x}{5} = 80\%$.

3. (5); $\frac{10 + k}{15 + k} = \frac{75}{100}$

$4(10 + k) = 3(15 + k)$

$40 + 4k = 45 + k$, so k = 5.

4. (D); There were p students present and 32 - p students absent.

$32 - p = \frac{x}{100} \bullet p$, and $x = \frac{100(32 - p)}{p}$.

5. (B); Let x = the total profit earned.

Then, $x - 0.2x = 1$

$0.8x = 1$, so x = \$1.20.

Practice Questions 5.9b

1. (16%); Use the increase/decrease formula to find the new price of the suit:

$k \bullet \frac{(100 - 40)}{100} \bullet \frac{(100 + 40)}{100} = k \bullet \frac{140}{100} \bullet \frac{60}{100}$

$= \frac{21k}{25}$.

$\frac{25 \text{ (original price)} - 21 \text{ (new price)}}{25 \text{ (original price)}}$

= 4/25, or 16%.

Therefore, there was a 16% decrease in the suit's price.

2. (72%); Let L = the length and W = the width of rectangle B.

Then, $Area_A = (0.9L)(0.8W)$

and $Area_B = (1L)(1W)$.

$\frac{Area_A}{Area_B} = \frac{(0.9)(0.8)}{(1)(1)} = \frac{0.72}{1} = 72\%$.

3. (43%); From the increase/decrease formula:

$x = \frac{(100 + 10)m}{100}$ and $y = \frac{(100 + 30)n}{100}$

Therefore, x = 1.1n and y = 1.3m,

$\frac{xy}{mn} = \frac{(1.1n)(1.3m)}{(1m)(1n)} = \frac{1.43}{1}$

Therefore, xy is 43% greater than mn.

4. (B); Let m = the marked price.

Then, $\frac{m(100 - 50)}{100} = \frac{x(100 + 10)}{100}$

$0.5m = 1.1x$, so $m = 2.2x$.

Practice Questions 5.10

1. (160 km/h); Let R = the speed of the kangaroo.

Then, using the formula RT = D,

R • 2.5 hours = 400 km

R = 160 km/h.

2. (3 hours); Let x = the total number of hours taken.

	R	T	D
Up	30 - 10	40/20 = 2	40
Down	30 + 10	40/40 = 1	40

The total time taken is 2 + 1 = 3 hours.

3. (5 p.m.); Let x = the number hours to overtake the slower man. Then, using the RT = D chart,

	R	T	D
Faster	40	x + 2	40(x - 2)
Slower	60	x	60x

Since both men travel the same distance,

$40(x + 2) = 60x$

$40x + 80 = 60x$

$20x = 80$, so x = 4 hours.

If the faster man leaves at 1 p.m., in 4 hours he will overtake the slower man at 5 p.m.

4. (10 miles); Let x = the distance from S to T. Then, using the RT = D chart,

	R	T	D
S to T	10	D/R = x/10	x
T to U	30	D/R = (40 - x)/30	40 - x

$\frac{x}{10} + \frac{(40 - x)}{30} = 2$

$3x + 40 - x = 60$

$2x = 10$, so $x = 10$.

Therefore, the distance from S to T is 10 miles.

5. (400 mph); Use the RT = D chart to find the time:

	R	T	D
1-way	600	D/R = 300/600 = 0.5	300
Return	300	D/R = 300/300 = 1	300

Average Speed = $\frac{\text{Total Distance } (300 + 300)}{\text{Total Time } (1 + 0.5)}$

= 600/1.5 = 400 mph.

Practice Questions 5.11

1. (5 nickels, 5 quarters); Let x = the number of nickels.

Then, 10 - x = the number of quarters.

$5x + 25(10 - x) = 150$

$5x + 250 - 25x = 150$

$20x = 100$

x = 5 nickels

10 - 5 = 5 quarters.

2. (70 cents/pound); The average price of a mixture is

$\frac{xP_1 + yP_2 + zP_3}{x + y + z} = \frac{3(30) + 2(150) + 2(50)}{7}$

= 490/7 = 70 cents.

3. (30 gallons); Let x = the number of gallons of water.

Since water is 0% solution,

$20(10) + x(0) = (20 + x)4$

$200 = 80 + 4x$

$4x = 120$, so x = 30 gallons.

4. (50/3 pints); Let x = the amount of pure sugar added.

$50(20) + x(100) = (50 + x)40$

$1{,}000 + 100x = 2{,}000 + 4x$

$60x = 1{,}000$

x = 50/3 pints.

5. (10/27); Let x = the portion of oil in the resulting solution. Since the oil in the original 10 kg solution is 2/5 of the solution, and the added 8 kg solution is 1/3 oil, the mixture formula in terms of the amount of oil is

$10(2/5) + 8(1/3) = 18(x)$
$60 + 40 = 270x$
$100 = 270x$
$x = 10/27$.

Practice Questions 5.12

1. (20/3 hours); Through the combined work approach,

$\left(\frac{1}{20} + \frac{1}{10}\right)x = 1$
$x + 2x = 20$
$3x = 20$, so $x = 20/3$ hours.

2. (9 hours); Through the multi-production approach,

(6 tractors)(12 hours) = (8 tractors)(x hours)
x = 9 hours.

3. (6 products); Through the multi-production approach,

$\frac{\text{(1 worker)(1 day)}}{\text{(1/2 product)}} = \frac{\text{(3 workers)(4 days)}}{\text{(x products)}}$
x = 6 products.

4. (3x/2 minutes); Let y = the number of minutes. Through the multi-production approach,

$\frac{\text{(1 machine)(x min.)}}{\text{(40 bolts)}} = \frac{\text{(2 machines)(y min.)}}{\text{(120 bolts)}}$
$3x = 2y$
$y = 3x/2$ minutes.

5. (7/3 hours); Let x = the number of hours John has to work alone to finish the job. Through the combined work approach,

$\left(\frac{1}{5} + \frac{1}{3}\right)1 + \left(\frac{1}{5}\right)x = 1$
$3 + 5 + 3x = 15$
$8 + 3x = 15$
$3x = 7$
$x = 7/3$ hours.

Chapter 5 Test

1. (C); $a + 3c = 2c + 1.5c$
$a + 3c = 3.5c$
$a = 0.5c$
$a/c = 0.5$.

2. (D); Since the numbers are consecutive integers,
$w = v + 2$
$x = v + 4$
$y = v + 6$
Then, $x + y = 2v + 10$, while $v + w = 2v + 2$.
$(2v + 10) - (2v + 2) = 8$.

3. (E); Let the number $T = 10t + u$. Then, by switching the variables we get $10u + t$.

$(10t + u) - (10u + t) = 9t - 9u$.
Therefore, the answer must be a multiple of 9, which 64 is not.

4. (A); Let x = Mary's current age. Then,

	Now	In 8 years
Mary	x	x + 8
Paul	2x	2x + 8

$n = 2x + 8$
$2x = n - 8$
$x = (n/2) - 4$.

5. (B); The total sum of the four test scores is 60(4) = 240. Adding the fifth test, the toal is then 240 + x, which is equivalent to the total sum of the five test scores:

$65(5) = 325$.
$240 + x = 325$, so $x = 85$.

6. (E); Let x = amount of gas needed to travel m miles.

Then, $\frac{g}{m + 1} = \frac{x}{m}$
$gm = x(m + 1)$
$x = \frac{gm}{m + 1}$.

7. (E); Movies V and S are sold in the ratio of 2:5, which means movie S was sold 5/7 of the time.

Then, $280 \bullet \frac{5}{7} = 200$ sales were of movie S.

8. (C); $\left(\frac{30}{100} \bullet 4x\right)\left(\frac{20}{100} \bullet x\right) = \frac{y}{100} \bullet (2x)^2$
$24x^2 = 4x^2y$
$y = 6\%$.

9. (D); Use the increase/decrease formula to find the net increase:

$\frac{n(100 - 10)}{100}\frac{(100 + 40)}{100} = \frac{126n}{100}$
Therefore, there will be a 26% net increase in n.

10. (B); Keeping in mind that the distance for both cases stays constant, create an RT = D chart to determine the new travel time for the train:

	R	T	D
Faster	55	4	55(4) = 220
Slower	50	220/50 = 4.4	220

4.4 - 4 = 0.4 more hours = 24 minutes.

11.(D); Let x = the fraction of the original solution that is salt. Using the mixture formula,

$1(100) + 1(0) + 2(x) = 4(50)$
$100 + 2x = 200$
$2x = 100$
$x = 50\%$.

12. (C); According to the given formula,
$2x + 2 = y$ and $2x + 2 = y$.
Therefore, $y = 10$ and $x = 4$, so $x + y = 14$.

13. (B); $6 + x = 10 - x$
$2x = 4$
$x = 2$.

14. (E); Let t = the smallest of the three integers.
Then, the next two integers are $t + 2$ and $t + 4$.
$t + t + 2 + t + 4 = x$
$3t + 6 = x$, so $t = (x - 6)/3$.

15. (D); The number is $10x + y$, so
$8(10x + y) = 80x + 80y$.

16. (C); Let x = the current age of the youngest sister.
The older sisters' ages will be $x + 2$ and $x + 4$.
Then, the sum of their ages two years ago:
$x - 2 + x + x + 2 = 30$, so $x = 10$.

17. (D); The sum of the three numbers is $50(4) = 200$.
So, $30 + m + n = 200$
The average of m and n is
$\frac{m + n}{2} = \frac{170}{2} = 85$.

18. (A); Through the proportion formula,
$\frac{t}{x} = \frac{y}{z}$, so $zt = xy$ and $z = \frac{xy}{t}$.

19. (E); Through the proportion formula,
$\frac{m}{m + n} = \frac{x}{60}$, so $x = \frac{60m}{m + n}$.

20. (D); Since 40% of x would be y, $0.4x = y$.
Then, $0.4x(4) = y(4) = 4y$.

21. (C); $v = 1.1x$ and $w = 1.2y$.
Then, $vw = (1.1)(1.2)xy$
$vw = 1.32xy$
Therefore, vw is 32% greater than xy.

22. (B); Create an RT = D chart to determine the total travel time:

	R	T	D
One-way	60	D/R = 600/60 = 10	600
Return	40	D/R = 600/40 = 15	600

Since the average speed is
$\frac{\text{Total distance } (600 + 600)}{\text{Total time taken } (10 + 15)} = \frac{1200}{25} = 48$ mph.

23. (D); Since pure water is 0% alcohol and pure alcohol is 100%, using the mixture formula:
$15(0) + 25(100) = 40x$
$2500 = 40x$
$x = 0.625 = 62.5\%$.

24. (36%); Let x = the percent of salt in the resulting solution. Since 2 liters of water evaporate,
$12(30) - 2(0) = 10x$
$360 = 10x$
$x = 36\%$.

Practice Questions 6.1

1. (68 pairs of shorts);

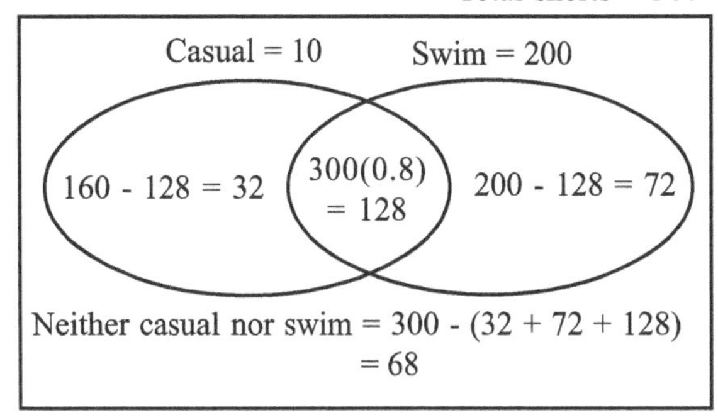

2. (9 students);

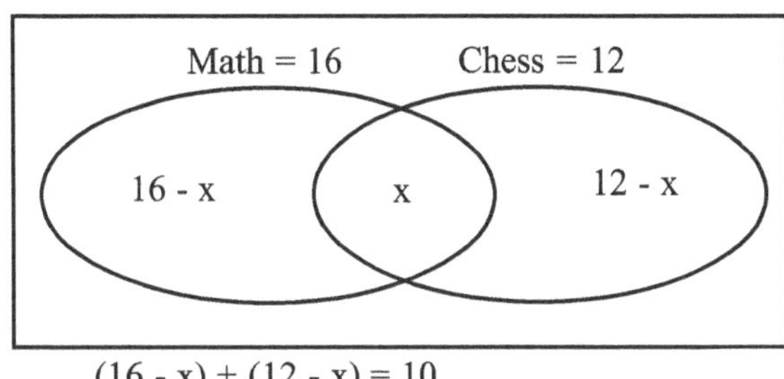

$(16 - x) + (12 - x) = 10$
$28 - 2x = 10$
$x = 9$ students.

3. (2 gift sets);

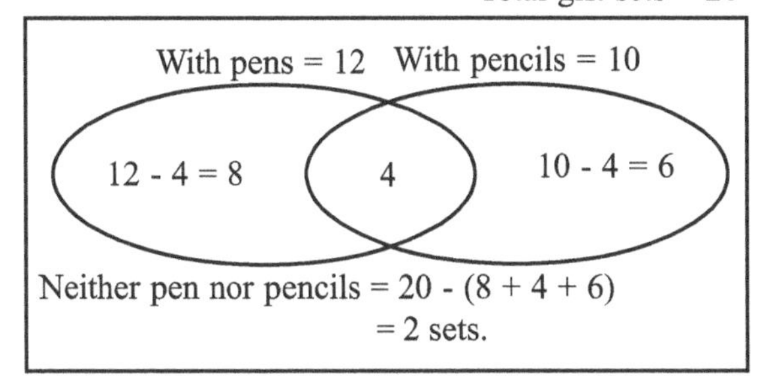

4. (37 customers);

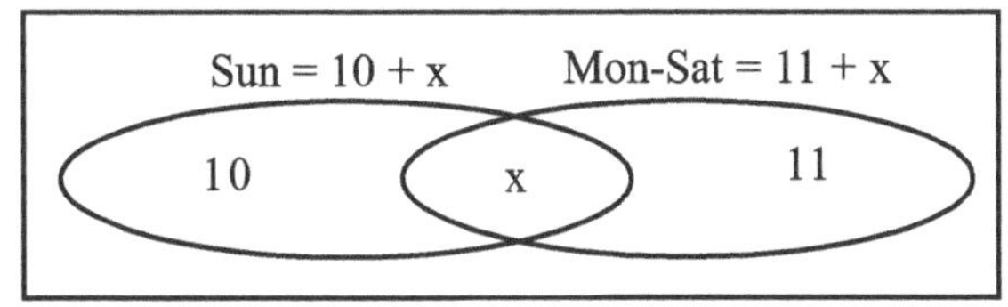

$10 + x + 11 = 48$
$x = 27$
$10 + x = 37$.

5. (120 students);

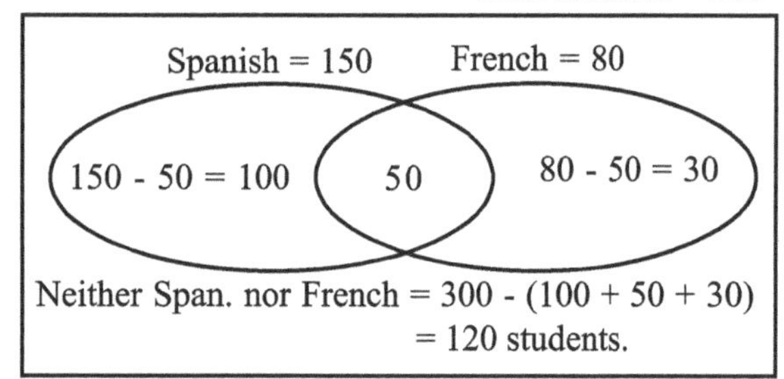

Practice Questions 6.2

1. (8 stamps); Susan receives
   From Marcy: 24(1/2)(1/3) = 4 stamps
   From Jane: 24(1-(1/2))(1/3) = 4 stamps.
   Therefore, Susan receives 4 + 4 = 8 stamps.

2. (120 tickets); Let x = the original number of tickets.
   Then, x - (1/2)x - 30 = (1/4)x
   4x - 2x - 120 = x
   2x - 120 = x, so x = 120 tickets.

3. (1/4); Since 2/3 of 3/4 of the workers are single males, 1/3 of 3/4 of the males workers are not single.
   (1/3)(3/4) = 1/4.

4. (150 students); Let x = the number of students in the university studying journalism.
   The number of students who receive a P, F, or W
   = (1/6)x + (1/5)x + (1/2)x = (26/30)x.
   Then, (4/30)x = the 20 remaining students (C),
   so x = 150.

5. (12 hamburgers); Let x = the original number of hamburgers in the refrigerator.
   Then, x - (1/6)x - (1/2)(5/6)x = 5.
   12 - 2x - 5x = 60
   5x = 60, so x = 12.

Practice Questions 6.3

1. ($2.90); Through the formula for total charge, the cost of a telephone call would be 1.50 + 0.20(x - 3), assuming that x is equal to or more than 3 minutes.
   Total charge = 1.50 + 0.20(10 - 3)
   = 1.50 + 0.20(7)
   = 1.50 + 1.40
   = $2.90.

2. (E); With the discount, the call is reduced to 20% of the original cost.
   Total charge = 0.2(1.20 + 5.2c) = 24 + 5.2c.

3. ($17.60); Let x = the total number of minutes. Assuming that x is greater than or equal to 1,
   Total charge = 0.50 + 0.30(x - 1)
   = 0.50 + 0.3(58 - 1) = 17.60.

4. (7/4 hours); Let x = the total number of quarter-hours parked. Then,
   140 + 100(x - 1) = 740
   40 + 100x = 740
   100x = 700, so x = 7.
   John paid for 7 quarter-hours, or 7/4 hours.

5. ($1,900); Since the truck was rented for 57 days, Total charge = 150 + 20(57 - 7) + 0.5(2,500 - 1,000).
   = 150 + 1,000 + 750
   = $1,900.

Practice Questions 6.4

1. (B); If cooked fish costs k/1, raw fish costs 4k/5. The difference, then, between buying raw insteasd of cooked fish is
   k/1 - 4k/5 = k/5, or 0.2k.

2. (E); Since the two containers are used equally in efficiency, we can equate the two values:

$$\frac{20}{x+3} = \frac{15}{x}$$

   20x = 15(x + 3)
   20x = 15x + 45
   5x = 45, so x = 9.

3. (B); Since what is being compared are the two unit prices, 4/5 versus 5/4,
   4/5 - 5/4 = 9/20, or $0.45.

Chapter 6 Test

1. (E); There are not enough given clues to solve the problem.

2. (D); x - (1/4)x - 1/2(3/4)x = (3/8)x of the sandwich is left.

3. (34 minutes); Let x = the number of minutes John spent on the phone.
   15 = 2.60 + 0.4(x - 3)
   1,500 = 260 + 40x - 120
   1,500 = 140 + 40x, so x = 34 minutes.

4. (C); In this percent decrease problem, remember the Bird's-Eye View strategy to break up the expression!

$$\frac{\frac{x}{300} - \frac{0.6x}{240}}{\frac{x}{300}} = \frac{\frac{x}{300}}{\frac{x}{300}} - \frac{\frac{0.6x}{240}}{\frac{x}{300}}$$

$$= 1 - \frac{(0.6x)(300)}{(240)(\ x\ )}$$

$$= 1 - 0.75$$

$$= 0.25 = 25\%.$$

5. (30 students);

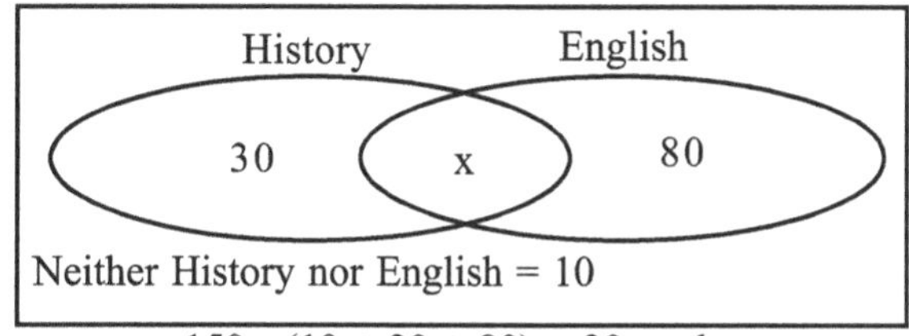

$x = 150 - (10 + 30 + 80) = 30$ students.

6. (A); Let x = the original amount of money.
$3 = x - 0.4x - 0.5(0.6x)$
$30 = 10x - 4x - 3x$
$3x = 30$, so $x = 10$
Then, $0.4x = \$4$.

7. (D); Since what is being compared is 25/2 and 5/1,
$25/2 - 5/1 = 12.5 - 5 = 7.5$ cents.

8. Let x = the total number of hours the truck was rented.
Then, in cents,
$1460 = 500 + 100(x - 2) - 20(x - 10)$
$1460 = 500 + 100x - 200 - 20x + 200$
$80x = 960$, so $x = 12$ hours.

Practice Questions 7.1

1. (y/2); Since the measure of an exterior angle of a triangle is the sum of the measures of its two remote interior angles,
$3x + 2y = 3y + x$
$2x = y$
$x = y/2$.

2. (C); The sum a triangle's exterior angles is 180, so
$x + 2x + 3y = 180$
$3x + 3y = 180$
$x + y = 60$.

3. (C); The sum of an n-gon's exterior angles is 360. Therefore, the sum of all the exterior angles of a decagon is 360.

4. (E); Since the sum of a triangle's exterior angles is the sum of its two remote interior angles,
$x = a + 45$ and
$y = b + 45$. Adding the two equations,
$x + y = a + b + 90$
$x + y = 120 + 90 = 200$.

Practice Questions 7.2

1. (B); Since M_1 and M_2 are not parallel,
I: d + f are supplementary = 180
II: e and g are corresponding angles but can never be equal
III: b and f are congruent corresponding angles.
Therefore, statement II cannot be true.

2. (E); Since ray AC bisects angle EAB,
$m\angle CAB = \frac{180 - 130}{2} = 25$.
Since angles BAD and ABF are supplementary,
$m\angle ABF = 180 - 130 = 50$.
Since ray BC bisects angle ABG,
$m\angle ABC = \frac{180 - 50}{2} = 65$.
Since the sum of the measures of the interior angles of a triangle is 180,
$m\angle ACB = 180 - 25 - 65 = 90$.

3. (E); Isolate one triangle and find the measures of all of its angles using the properties of parallel lines. In this case, use the uppermost one:
Since $x + y = 2z$, $x + y + (180 - 2z) = 180$
$x + y - 2z = 0$, so $x = y - 2z$.

4. (D); If lines L_1 and L_2 are not parallel, e and c could not be equal. However, since lines L_1 and L_2 are parallel, e and c could be equal.

Practice Questions 7.3

1. (D); Since AB = AC and BD = DC, we can conclude that $m\angle ADB = 90$. Then, $m\angle ABD + m\angle BAD = 90$.

2. (E); Since angle ABD is not necessarily a right angle, angles p and s do not necessarily add up to 90.

3. (E); $m\angle ACB = 180 - 135 = 45$.
Since lines BA = BC, $m\angle ABC = m\angle BCA = 45$, which means the triangle is isosceles.
However, since $m\angle BAC = 180 - 2(45) = 90$, the triangle is a 45-45-90 triangle, which cannot be an equilateral triangle.

4. (D); $x = 2y - 30$, so $x + y = 90$, which is the sum of the two acute angles of a right angle.
$2y - 30 + y = 90$
$3y = 120$, so $y = 40$.
$x + y = 90$, so $x = 50$.

5. (C); The sum of the measures of the interior angles of both triangles ABC and ABD = 180, so we can equate them:
$3x + y = 2y + x$
$2x = y$
$x = y/2$
Then, since $2y + x = 180$
$2y + y/2 = 180$
$5y/2 = 180$, so $y = 72$.
Finally, $x = y/2 = 72/2 = 36$.

Practice Questions 7.4

1. (D); Draw a diagonal to separate the figure into two right triangles. Then, the triangle to the right consists of the Pythagorian triple 7:24:25, while the triangle to the left consists of the Pythagorian triple 3:4:5. Therefore, 3:4:5 = 15:20:25, so x = 15.

2. (10); The two triangles are made up of two Pythagorean triples, 8:15:17 on the left and 6:8:10 on the right. Therefore, x = 10.

3. (C); Let L = length and W = width of the rectangle.
Then, $2L/3 = 4W/3$
$W = L/2$, so $L = 2W$.
Using the Pythagorean theorem,
$(W)^2 + (L)^2 = D^2$
$(W)^2 + (2W)^2 = D^2$
$D^2 = 5W^2$, so $D = w\sqrt{5}$.

4. ($2\sqrt{2}$); Since the triangle to the right is a 30-60-90 triangle, we can determine that the length of its other leg is 2. Since that leg is shared with the triangle to the left, which is a 45-45-90 triangle, we can determine that its hypotenuse is $2\sqrt{2}$.

5. (60); Through the Pythagorean theorem, we determine that the sides of the triangle are 1, $\sqrt{3}$, and 2. Therefore, it is a 30-60-90 triangle, with AB = 1. $m\angle BAC = 60$.

Practice Questions 7.5

1. (E); $AB - AC < BC < AB + AC$
$8 - 3 < BC < 8 + 3$
$5 < BC < 11$
Therefore, BC has to be greater than 5 or less than 11.

2. (A); $17 - 5 < x < 17 + 5$
$12 < x < 22$
Side x is the longest, which means it must be greater than 17. Then only prime number between 17 and 22 is 19.

3. (B); According to the Scissors theorem, when two adjacent sides of a triangle are the same between two triangles, the third side is greater if its opposite included angle is larger. Since both triangles have the same sides, k > h because 30 > 29.

4. (E); Since the triangle has sides of different lengths, it cannot be equilateral and its longest side is t + 2. Therefore, the largest angle must be greater than 60.

Practice Questions 7.6

1. (B); The circumference of a circle is πd, which in this case, is πs.

2. (C); $2(2x - 3) + 2(x + 3) = 60$
$4x - 6 + 2x + 6 = 60$
$6x = 60$, so $x = 10$.
Then, $2x - 3 = 2(17) - 3 = 17$.

3. (B); While Fig. 1 has side s, Fig. 2 has altitude s, which means its side (the hypotenuse of the formed triangle) is longer. Therefore, Fig. 2's perimeter is longer than Fig. 1's.

4. (C); The circumference of the wheel is
$2\pi r = 2(22/7)(14) = 88$.
$\frac{1200 \text{ inches}}{88 \text{ inches.}} = 14$ revolutions.

5. (C); Let L = the length and W = the width.
$2L + 2W = 3L$
$L = 2W$
$3L = 6W$.

6. (B); Let x = the number of meters of additional bricks needed on one side. Note that we want to extend the 20m side as opposed to the 40m side to have the maximum increase in area for the minimum number of bricks.
$40(20 + x) = 820$
$800 + 40x = 820$
$40x = 20$, so $x = 0.5$
Since two sides are needed, the answer is 2(0.5) = 1 meter.

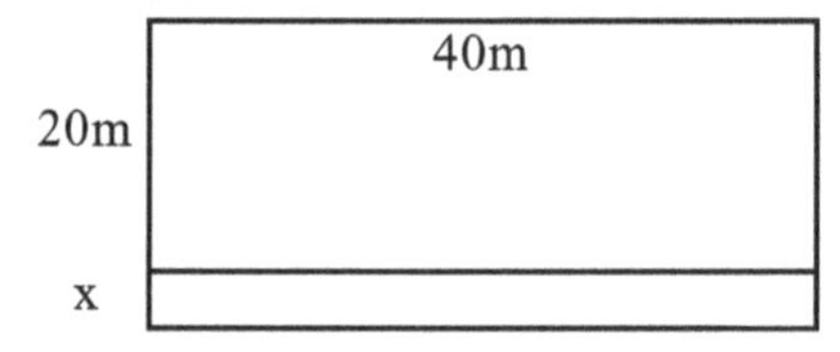

Practice Questions 7.7

1. (E); If you draw altitude AD to side BC, you create a 30-60-90 triangle, with AD being $2\sqrt{3}$. Therefore, the area of the triangle is $(1/2)bh = (1/2)(60(2\sqrt{3}) = 6\sqrt{3}$.

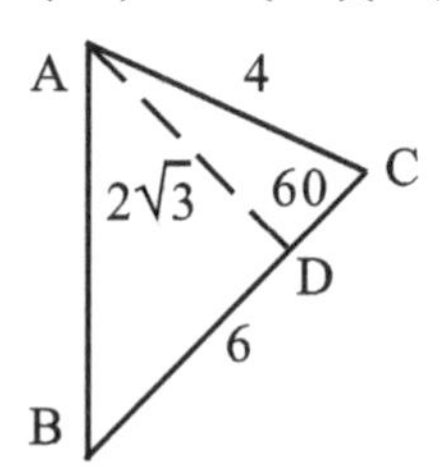

2. (B); If the square is inscribed in the circle, the diagonal of the square is the diameter of the circle.
$4\pi = \pi r^2$
r = 2, so d = 4.
Use the Pythagorean theorem to find the side of the square:
$x^2 + x^2 = 4^2$
$2x^2 = 16$
$x^2 = 8$.
The area of the square is x^2, so the answer is 8.

3. (A); Since the radius of the semicircle is 1, the diameter and thus the side of the triangle is 2.
Then, the area of the equilateral triangle is
$\frac{s^2\sqrt{3}}{4} = \frac{2^2\sqrt{3}}{4} = \sqrt{3}$.

4. (B); Since the area of a triangle = (1/2)bh,
$32 = (1/2)x^2$
$x^2 = 64$, so x = 8.
Since the isosceles right triangle is a 45-45-90 triangle, hypotenuse is $8\sqrt{2}$.

Practice Questions 7.8

1. (D); This problem is best solved by putting the shaded pieces together as in a jigsaw puzzle.

2. (B); Since the two figures retain essentially the same dimensions, the pereimeters are equal.

3. (C); To figure out the perimeter, we only need to find the length of half of the figure. Since the two known sides have lengths of 11, the perimeter is 44.

4. (C); Use the shadow technique and subtract the areas of the non-shaded regions to find the area of the shaded region. As shown below, we could draw a vertical line to create a smaller rectangular region that surrounds the shaded triangle:

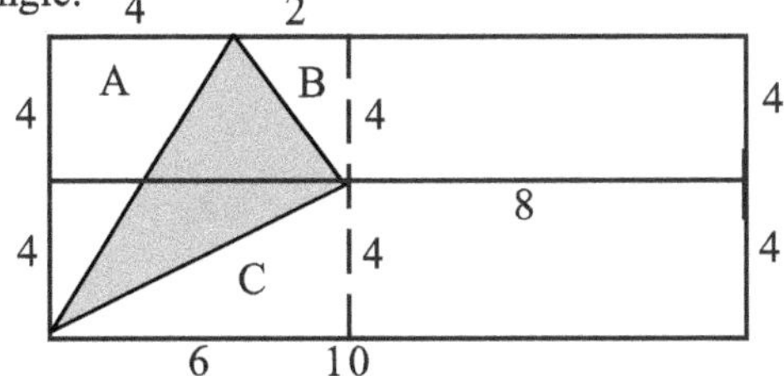

Area of smaller rectangle = (6)(4) = 24
$Area_A = (1/2)bh = (1/2)(4)(4) = 8$
$Area_B = 4$
$Area_C = 12$
Area of shaded triangle = 24 - 8 - 4 - 12 = 16.

5. (B); Triangle ABC is divided up into four triangles of equal areas: ADF, CFE, BDE, and DFE. Triangle DFE is divided in half, making the area of triangle DFG 1/8 of the the area of ABC, which is 2.

Practice Questions 7.9

1. (C); If the radius of the circle is 10,
the circumference is $2(\pi)(10) = 20\pi$.
The length of the arc PQ is 5π, which is 1/4 of the circumference.
Then, drawing chord to connect the center of the circle to points P and Q will create a 45-45-90 right triangle, made up of sides of length 10 and a hypotenuse of length $10\sqrt{2}$.

2. (E); If the four resulting chords AB, BC, CD, and DA were of equal length, the degree measures of arcs AB, BC, CD, and DA would all be 360/4 = 90. However, since chord AB is the longest, the degree measure of arc AB must be greater than 90.

3. (D); We can draw a right triangle using the radius of the larger circle, the radius of the smaller circle, and half the side of one side of the triangle. Then, using the Pythagorean triple of 3:4:5, we can determine that the length of one side of the equilateral triangle is 4(2) = 8, which means the perimeter will be 8(3) = 24.

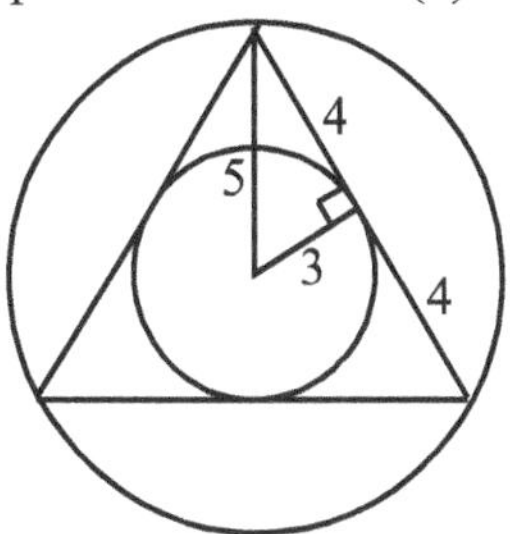

4. (C); Let x = the radius of the circle. Then, since the length of the two exterior tangent lines drawn from an outside point to a circle are equal, the other sides of the triangle are 3 - x and 4 - x.
$2(3 - x) + 2(4 - x) + 2x = 12$
$6 - 2x + 8 - 2x + 2x = 12$
$14 - 2x = 12$
2x = 2, so x = 1.

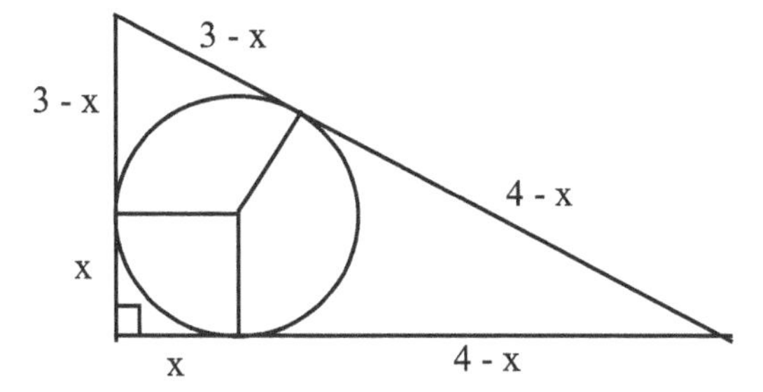

5. (C); Since the arc EF is a quarter circle, the circumference of the circle is $4(12\pi) = 48\pi$
Then, to find the diameter of the circle,
$48\pi = \pi D$, so D = 48.
Then, the radius is D/2 = 24.

Practice Questions 7.10

1. (B); Since there are six sides to a cube, the surface area of the cube is $0.24 = 6s^2$.
Then, $s^2 = 0.04$, and s = 0.02
The volume of the cube = $s^3 = (0.02)^2 = 0.008$.

2. (C); The volume of the rectangular tank is 4 • 7 • 11 = 308, while the volume of the cylindrical tank is $(\pi)(3.5^2)h$. Equating the two volumes,
$308 = (22/7)(3.5^2)h$
$h = \frac{308(7/22)}{12.25} = \frac{98}{12.25} = 8.$

3. (B); $d^2 = 2^2 + 1^2 + h^2$
$d^2 = 4 + 1 + h$
$d^2 = 5 + h^2$.

4. (C); Area ratios are length ratios (in this case, circumference ratios,) squared. Therefore, 5 : 2 squared is 25 : 4. However, the shaded area is found by subtracting the area of the smaller circle from the area of the larger circle.
Therefore, the area ratio is (25 - 4) : 4 = 21 : 4.

5. (C); Area ratios are length ratios squared, and volume ratios are length ratios cubed. Given the area ratio between the bases of the two similar cones,
$\frac{a^2}{b^2} = \frac{9}{4}$, so $\frac{a}{b} = \frac{3}{2}$.
Then, $V = \frac{a^3k}{b^3k} = \frac{3^3k}{2^3k} = \frac{27k}{8k}$
Therefore, the volume of the smaller cone is 8k.

Practice Questions 7.11

1. (E); From the figure we can deduce that the radius of the circle is 3. The circle equation, then, is $x^2 + y^2 = r^2$, and in this case, $x^2 + y^2 = 3^2 = 9$.
Plugging the coordinates into the equation, $(-2)^2 + (-\sqrt{5})^2 = 4 + 5 = 9$. Therefore, the circle passes through the point $(-2, -\sqrt{5})$.

2. (E); According to the number line, A/D is approximately 1/3. C/G, then, has the most similar ratio of approximately 1.3/1.9 = 1.3.

3. (C); k - 8 = 6 - 4, so k = 10
t - 3 = 10 - 3, so t = 10.
Therefore, the coordinate B is (10,10).

4. (A); By plotting all of the answer choices, one can find that only (14,7) lies outside of the triangle.

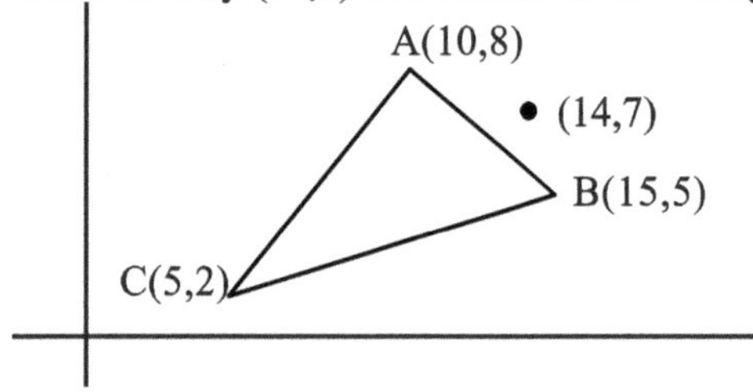

Practice Questions 7.12

1. (A); The maximum area of a polygon is found when the sides multiplied together, (in this case, x and y,) are as close to each other in length as possible. Then, if x = 8 and y = 7, 8 • 7 = 56.

2. (D); There are 4 separate parallelograms and 5 more overlapping ones, for a total of 9.

3. (C);

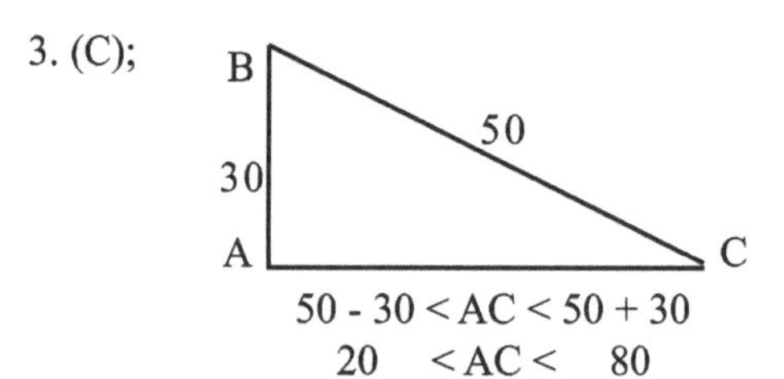

$50 - 30 < AC < 50 + 30$
$20 < AC < 80$
Therefore, the closest minimum is 21 and the closest maximum is 79.

4. (B); When perimeters are constant, the polygon with more sides will have a greater maximum possible area.

5. (D); Draw a picture for this kind of problem. An example for the maximum number of non-overlapping triangles is represented in the following diagram.

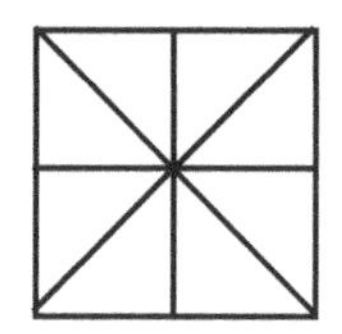

6. (C); The shortest route consists of 4 possible routes from A to R and 10 possible routes from R to T. Therefore, there are (10)(4) = 40 possible routes.

Practice Questions 7.13

1. (C); $2\pi r = 2/\pi$, so $r = 1/\pi^2$.
Then, the area $\pi r^2 = \pi(1/\pi)^2 = 1/\pi$.

2. (C); $V = s^3 = 2h$
Then, $6s^2 = 3h$, so $s^2 = h/2$.
$\frac{s^3}{s^2} = \frac{2h}{h/2} = 4$.

3. (A); 140 - 3x + x + y = 180
y - 2x = 40
y = 2x + 40
Therefore, $40 < 2x + 40 < 60$
$0 < 2x < 20$
$0 < x < 10$.

4. (C); $\frac{xy}{xy + (1 - xy)} \bullet 100\% = \frac{xy}{1} \bullet 100\% = 100xy\%$.

5. (D); The area of of the square is h^2. Since the shaded area is 1/3 of the non-shaded area, it is also 1/4 of the total area and half of the area of triangle ADC. Therefore, the area ratio between the shaded triangle EDF and triangle ADC = 1 : 2. Since the length ratio between similar triangles is the square root of their area ratio, let x = the length of EF:
$\frac{x}{AC} = \frac{\sqrt{1}}{\sqrt{2}}$, so $x = \frac{AC}{\sqrt{2}}$.
Since $AC = h\sqrt{2}$ (45-45-90 triangle rule),
$x = \frac{AC}{\sqrt{2}} = \frac{h\sqrt{2}}{\sqrt{2}} = h$.

Practice Questions 7.14

1. (C); If the small segment along the chord of length $4\sqrt{2}$ is cut off, the remaining segment will be composed of 3/4 the circle and a triangle with the area

$(1/2)bh = (1/2)(4)(4) = 8$:

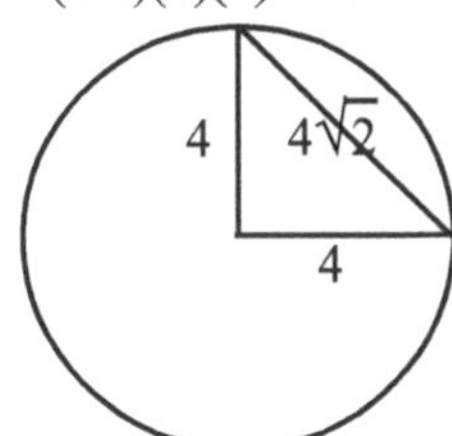

Therefore, the area is $(3/4)\pi(4^2) + 8 = 12\pi + 8$.

2. (C); An edge of the triangle is $4\sqrt{2}$ (45-45-90 triangle), so the area of an equilateral triangle, which has the maximum possible area, is

$$\frac{s^2\sqrt{3}}{4} = \frac{(4\sqrt{2})^2(\sqrt{3})}{4} = \frac{32\sqrt{3}}{4} = 8\sqrt{3}.$$

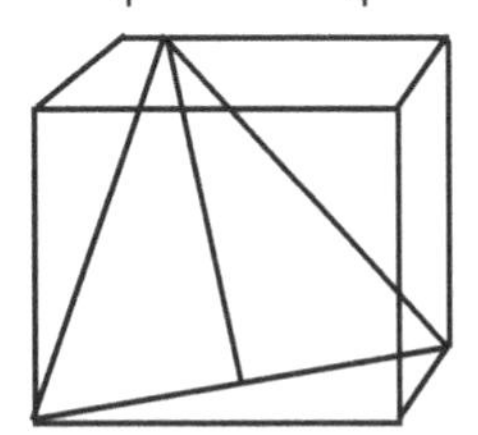

3. (C); All of the choices except C can be false:

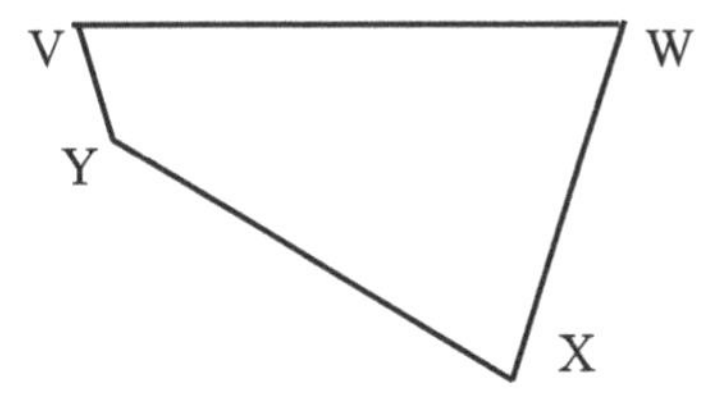

4. (3/4); Draw a diagram according to the given specifications.

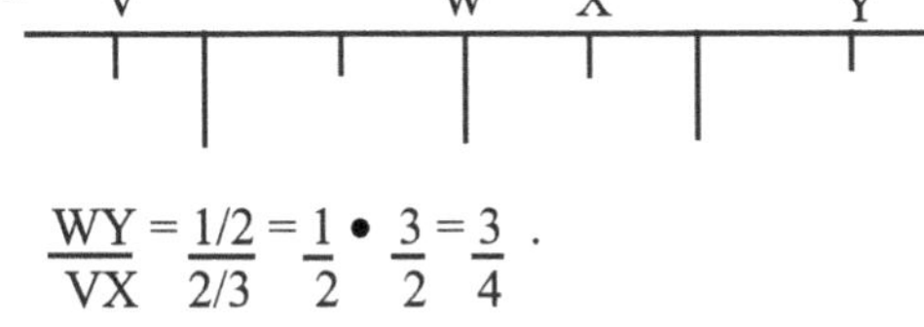

$$\frac{WY}{VX} = \frac{1/2}{2/3} = \frac{1}{2} \bullet \frac{3}{2} = \frac{3}{4}.$$

Chapter 7 Test

1. (B); I: x and k are supplementary angles.
II: $180 - (h + m) = k$
II: the measure of an exterior angle of a triangle is the sum of the measures of its two remote interior angles.
Therefore, statement II is not true.

2. (E); Statement I is only true if M_1 and M_2 are parallel, in which case a and b would be supplementary angles.

3. (A); A: ABC is not necessarily a right triangle because the measure of angle A is unknown.
B: $\angle CAD$ and $\angle ACD$ are equal, so the triangle must be isosceles.
C: If $\angle CAD$ and $\angle ACD$ are equal and $\angle ADC = 90$, $\angle CAD$ and $\angle ACD$ would each $= 45$.
D: $\angle ADC$ is a right angle.
E: $\angle x$ and $\angle y$ are complementary angles.

4. (D); Triangle ABC is a 30-60-90 triangle. Triangle BDC is also a 30-60-90 triangle. Therefore, the length of BD is x/2 and the length of AD is $\frac{x/2}{\sqrt{3}} = \frac{x\sqrt{3}}{6}$.

5. (B); $7 - 3 < x < 7 + 3$
$4 < x < 10$.
Therefore, x can be 5, 6, 7, 8, and 9.

6. (E); The circumference of the wheel is $2k\pi$. Then, the number of revolutions is $\frac{500}{2k\pi} = \frac{250}{k\pi}$.

7. (E); This is a 45-45-90 triangle, so two sides are the same. Through the area formula,
$A = (1/2)bh$
$32 = (1/2)(x^2)$
$x = 8$.
Therefore, the first two sides are 8 and the hypotenuse is $8\sqrt{2}$.
The perimeter is $8 + 8 + 8\sqrt{2} = 8 + 16\sqrt{2}$.

8.(C); Through the Bird's-Eye View method in perimeter calculations, we can see that the perimeter of regions C + A is the sum two radii and the circumference of a quarter circle.
The diagonal of the square is equal to the radius of the circle: since the area of the square is x^2, each side is x and the diagonal is thus $x\sqrt{2}$ (45-45-90 triangle). Therefore, the sum of two radii is $x\sqrt{2} + x\sqrt{2} = 2x\sqrt{2}$.
The perimeter of the quarter circle, then, is

$$\frac{2\pi r}{4} = \frac{2(\pi)(x\sqrt{2})}{4} = \frac{\pi x\sqrt{2}}{2}.$$

Therefore, the sum of the two radii and the circumference of the quarter circle is

$$2x\sqrt{2} + \frac{\pi x\sqrt{2}}{2} = x\sqrt{2}(2 + \frac{\pi}{2}).$$

9. (D); Let x = the length of chord CD. Since arc AB is 90°, we know the triangle formed by the center of the circle, the middle of AB, and point B is a 45-45-90 triangle. Then, we can deduce that the leg of the triangle is 1, while the other is x/2. (See diagram on following page.

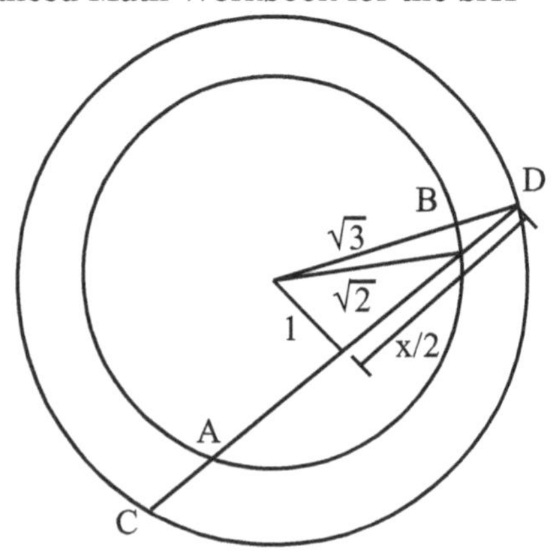

From the Pythagorean theorem:
$12 + (1/x)^2 = (\sqrt{3})^2$
$1 + (1/4)x^2 = 3$
$(1/4)x^2 = 2$
$x^2 = 8$
$x = 2\sqrt{2}$.

10. (D); Area ratios are length ratios squared, while volume ratios are area ratios cubed. Therefore, if the length ratio is 2 : 3, the volume ratio is $2^3 : 3^3 = 8 : 27$, and since the volume of the smaller sphere is 8(2) = 16, the volume of the larger sphere is 27(2) = 54.

11. (D); The figure below is an example of how four lines can divide an equilateral triangle into 11 non-overlapping regions:

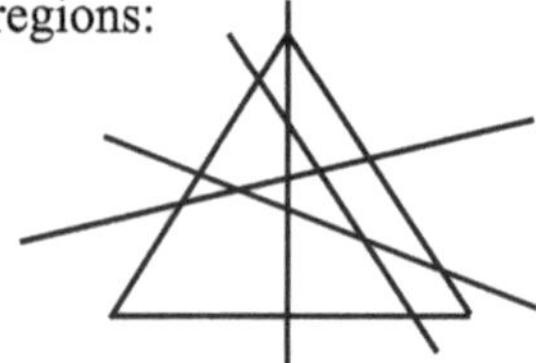

13. (E); Let s = the length of an edge of the cube. Since the diagonal of a cube is $\sqrt{3}$ times the length of an edge,
$3s^2 = 9$
$s^2 = 3$
$s = \sqrt{3}$
Then, the surface area = $6s^2$
$= 6(\sqrt{3})^2$
= 18.

14. (C); Altitude AD is longer than altitude BE because angle B is greater than angle A and side AC is thus greater than side BC.

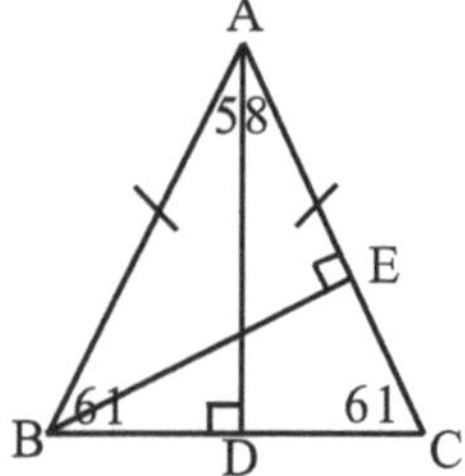

CPSIA information can be obtained
at www.ICGtesting.com
Printed in the USA
LVHW061541110122
708310LV00012B/1252